THE STORY OF ASTRONOMY

Peter Aughton is the author of ten acclaimed popular history titles including *Endeavour*, *Resolution* and *Voyages that Changed the World*, as well as biographies of Jeremiah Horrocks and Isaac Newton. Formerly a computer engineer in the aerospace industry, where he worked on the world's first supersonic airliner, he went on to lecture at the University of the West of England for 25 years.

THE STORY
OF ASTRONOMY

Peter Aughton

Quercus

First published in Great Britain in 2008 by Quercus

This paperback edition published in 2011 by
Quercus
55 Baker Street
7th Floor, South Block
London
W1U 8EW

A CIP catalogue record for this book is available
from the British Library.

ISBN 978 0 85738 598 7

10 9 8 7 6 5 4 3 2 1

Text designed and typeset by Ellipsis Digital Limited, Glasgow
Printed and bound in Great Britain by Clays Ltd, St Ives plc

CONTENTS

INTRODUCTION   1

1 THE ORIGINS OF ASTRONOMY   5

2 FROM BABYLON TO ANCIENT GREECE   16

3 THE ALMAGEST   35

4 ASTRONOMY IN THE DARK AGES   51

5 THE COPERNICAN REVOLUTION   67

6 CHASING THE PATHS OF THE PLANETS   81

7 GALILEO
  The Great Telescope Maker   96

8 JEREMIAH HORROCKS
  Father of English Astronomy   109

9 THE CLOCKWORK UNIVERSE   128

10 ENGLISH AND FRENCH RIVALRY   147

11 FINDING LONGITUDE   161

12 WILLIAM HERSCHEL
   Gazing Deeper into Space   178

13 UNDERSTANDING THE FORCES OF
NATURE   192

14 ALBERT EINSTEIN
Relativity Redefines Astronomy   210

15 THE HUBBLE UNIVERSE   228

16 FROM MICROCOSM TO MACROCOSM   244

17 BEYOND THE VISIBLE SPECTRUM   259

18 BLACK HOLES, QUASARS AND
THE UNIVERSE   276

19 STEPHEN HAWKING
Exploring the Boundaries of Space   290

20 ASTRONOMY IN THE SPACE AGE   305

21 THE BIG BANG AND THE CREATION
OF THE UNIVERSE   326

22 DARK MATTER AND DARK ENERGY   340

23 PLANETS, MOONS AND THE SEARCH
FOR LIFE   358

GLOSSARY   367

FURTHER READING   379

INDEX   381

INTRODUCTION

From the time of the very earliest civilizations people have wondered about the world they live in, about how it was created and about how it will end. Not the least they wanted to understand the mysterious world of the heavens above them. In this book we travel on a journey from those earliest peoples as they stared in fear and wonder at the night sky to the 21st century cosmologists as they peer into the deepest corners of space, often looking back in time.

Our journey follows an essentially chronological sequence. It pauses to look deeper at the significant discoveries and world of famous astronomers and scientists (such as Copernicus, Galileo, Newton and Hubble). The story is also punctuated by descriptions of significant advances in technology, such as that of the light telescope or space telescope, which have enabled astronomers to make huge leaps forward in their understanding of the universe. The role of the mathematicians and other theorists (from Aristarchus to Einstein and Hawking) are not forgotten –

at many stages in this story they have propelled our under-standing of the cosmos to new levels.

In the most distant times the Sun was seen to make its daily journey across the sky, followed each night by the arrival of the Moon. The Moon was seen to wax or wane a little each time it appeared, and on a few nights it did not appear at all. Tiny specks of light dotted the great black dome of the heavens; some of these appeared in different nightly positions against their neighbours and became known as planets. It took millennia for man to determine the true nature of these wandering stars and to evolve a model of the world to accommodate them and to predict their positions in the sky.

It was thought that every star had its own purpose and that the secrets of the universe could be discovered by making a study of them. The telescope was invented and the hundreds of stars became thousands and then tens of thousands. There was the telescope of Galileo, followed by observatories at Paris and Greenwich, the telescopes of William Herschel and later the giant tele-scopes of Mount Wilson and Mount Palomar. At the end of the second millennium came the Hubble Space Tele-scope. The number of stars in the catalogues grew to hundreds of thousands and the number of stars visible in the telescopes became millions, and then billions. The number of stars in the sky became countless. How could these tiny points of light in the sky explain the beginnings of the world?

The tiny dots of light came under more and more scrutiny. Many of them were not simple dots at all, but whole galaxies, each consisting of millions or billions of stars. It was discovered that there was much more to light than first appeared. A star made an image on a photographic plate, but the light from the star could be spread out into its different wavelengths by passing it through a prism. The result was amazing – not only could astronomers discover the very elements the stars were made of but they could also work out the speed at which a star was approaching us or rushing away from us. There was more: some stars did not have a constant brightness and their distance could be calculated as they went through a cycle of variable brightness.

Then man discovered other wavelengths of radiation beyond the visible spectrum. The stars emitted not only visible light but also infrared, ultraviolet, radio frequencies, microwaves, gamma rays and X-rays – much more was discovered about the stars by observing them in these wavelengths. The astronomer had to call on the physicist to explain many of his results; upon the mathematician to work out theories of gravitation and relativity; and upon the engineer to build more powerful telescopes, eventually sending his instruments into orbit to escape the filtering effects of the atmosphere.

Starlight provides all the information we need to put together the story of how we come to be here, on what may well prove to be a unique planet in the universe. We still do not know the answers to all of the questions, but we can now build on the observations of hundreds of generations of astronomers. At the start of the third millennium, we can put together a credible account of the story of the creation of the universe.

Peter Aughton
2008

1

THE ORIGINS OF ASTRONOMY

To the earliest peoples the heavenly bodies were a source of fear and wonder. The rising and setting Sun and the waxing and waning Moon seemed to be things controlled by the gods. In time, civilizations began to link celestial sights with earthly events, noting that they often coincided with fertile floods, harvests and other important features of everyday life. Later, they began to plot the positions of the stars and planets, and to give them names and meanings. So began the first studies in astronomy and astrology. By observing the heavens, the first calendars could also be made.

On the appointed day, in broad daylight, the whole community gathered outside the temple. There was a hush of reverence as the astronomer priests conducted a solemn ceremony before the frightened and apprehensive crowd. The astronomer priests had calculated that the Moon was rapidly approaching the Sun and that a once-in-a-lifetime event was about to happen. The wise refrained from looking at the Sun, but the foolish stared and saw that a small dark bite had been taken from the shining golden disc. At first the bite was trivial in size, but as it grew steadily it became a large, dark shadow blocking out the sunlight. Soon it was obvious even to those who lowered their eyes that the Sun had become weaker. Daylight was fading. There was a fear amongst the crowd that the Sun was under attack from an unknown and powerful enemy. Not even the village elders could remember an event like this before. What was this unknown force that was draining the strength of the Sun? What if the Sun were to be extinguished altogether and the world doomed to live in a state of perpetual darkness?

Now the Sun was covered completely. It became dark – not totally black, but as if it were no longer day-time. There was a great hush as the community waited in fear. The birds had been chattering, but like the people present they, too, suddenly fell silent. Never had there been such a breathtaking silence when so many had been gathered together. Those who dared still gazed at the sky. They saw that the stars and planets, the familiar patterns of light in the night sky, had appeared in the middle of the day. These were sights that belonged with the night, not the daytime. It was a terrible omen from the gods. The total darkness lasted only a few minutes but it seemed much longer, and in the middle of the day it felt much stranger than any darkness they had known before.

Some believed the end of the world had come. They waited hushed in the darkness and they stared at the stars. Then, after a time, there came a small but brilliant flash of light from the edge of the dark disc. It became a little brighter and then it appeared like a string of golden beads as the craters and the mountains of the Moon broke the thin crescent of sunlight into an irregular arc of fire. There was a murmur of expectation, and the arc of light grew brighter until the slender crescent could be seen behind the black disc that many thought was an evil dragon consuming the Sun. But the astronomers knew that the dragon in the sky was the familiar Moon. The crescent of light grew slowly, and by degrees the Sun began to appear again. There was a great feeling of relief. Perhaps the Sun had not been destroyed after all. Then the silence was broken and everybody was chattering about the event. They had seen a spectacle that they would never forget. Many years later the children would tell the tale to their grandchildren, but those who were not present would never believe the details of their story.

Observing and Understanding

Of all the branches of knowledge known to us, astronomy has always held a special place. It has a history that spans every era of human development, and its origins can be traced back to the earliest civilizations. It holds a special place because it strives to explain the origins and the purpose of the universe. Every new attempt to study and measure the skies reveals new and surprising facts. As soon as one astronomical feature appears to be solved and understood, new and stranger phenomena are discovered.

The Sun and the Moon are the most prominent objects in the sky; it is pure coincidence that they appear almost the same size to us. But there is also a myriad of tiny dots of light in the night sky, and these, too, must be studied to discover the secrets of the nature and origins of the universe. The deeper we penetrate into the night sky and the more we study the objects we find, the deeper becomes the mystery of creation – a mystery that we will probably never solve to our complete satisfaction.

In the ancient world, as the first stable civilizations developed, there were always some who were fascinated by the night sky. The shepherd in his lonely vigil had plenty of time to study the sky, to learn the positions of the stars, to observe the phases of the Moon and to notice that the stars appeared in the same patterns every night. Whilst the stars maintained these patterns, the Sun and the Moon had their own independent motion across them. The whole night sky appeared to revolve throughout the night about a point in the north; it was as though the sky were painted on the inside of a huge globe rotating about the Earth. There were also seasonal differences in the sky. The night sky in winter did not have the same appearance as the night sky in summer. Sometimes new constellations (patterns of stars) appeared near the horizon and old constellations disappeared. There was an annual cycle, and the missing constellations always returned as the seasons changed. There was also a small number of star-like objects that wandered across the night skies; they sometimes exhibited a backwards (or retrograde) motion before progressing forwards once more.

Every civilization wondered about the heavens. Priesthoods developed in which men were trained to make a special study of the stars. The Sun was very obviously the most influential object in the sky. It was so strong, bright and powerful that as soon as it rose in the morning all the stars faded from view. The purpose of the Sun was obvious. It gave light in the daytime and it helped the crops to grow. The purpose of the Moon was not so obvious; although it often gave a pale light by night, there were many nights when it only made a very late appearance and there were nights when it gave out no light at all. It exhibited phases, from the thin crescent of the new Moon to the bright circle of the full Moon, with every other possible phase in between. These phases were easily predicted, and it was not difficult to deduce that they depended on the relative position of the Sun. It did not take astronomers very long to realize that the Moon was actually a sphere, and the phases came about because it was illuminated by light from the Sun. It came as a revelation when it was found that the phases of the Moon were also related to the movement of the tides. The Moon obviously held a great sway over the seas, but it was hard to explain in physical terms how a sphere travelling across the sky in a monthly cycle could affect the tides. It was logical to deduce that every object in the skies had some small influence on the Earth. The learned people of the day worked hard to discover the nature of these influences.

Discovering the Planets

As we have noted, there were also other objects in the heavens that were visible to early astronomers apart from the Sun, the Moon and the stars. There were wandering 'stars' that seemed to have their own motion. It was thought that these bodies, which became known as the planets, must also have some strange influence over the Earth. One of these planets was Venus, the 'morning star' that sometimes heralded the dawn, but which at other times appeared in the evening when it was known as the 'evening star'. It was the brightest planet in the sky. There was the red planet that became known in the Roman world as Mars. There were the slower-moving planets that the Romans called Jupiter and Saturn. There was a small planet they called Mercury that was difficult to see, for it stayed close to the Sun. It was assumed that the planets had all been put in the sky for a reason, and that every star had a message to tell. These reasons and messages were not easy to explain, however.

Signs from the Heavens

In Egypt, it was discovered that the rising above the horizon of Sirius, the Dog Star, heralded the flooding of the River Nile. It was a great triumph for the astronomer priests to be able to predict when the Nile was about to flood, for it enabled the farmers to prepare themselves for the event – fertile soil brought down by the river enriched the farmland and improved the harvest. The astronomer priests also hoped to discover the purpose of other stars in the sky. The occasional meteor flashed across the heavens. It never returned, but it was not difficult to make a link with the phenomenon and some earthly event – an important death or a birth, for example. Less frequently a comet appeared; this was so rare that it was frequently deemed to be a bad omen in many cultures. Surely, reasoned the astronomer priests, all these portents in the skies were signs from the gods trying to tell them something.

The development of astronomy and astrology were not confined to Egypt and the Mediterranean world. The Chinese, the Aztecs and the Incas all studied the stars and tried to predict eclipses and other events in the heavens. Some of the stars appeared to form distinct constellations, and the vivid imagination of the ancient astronomers saw all kinds of patterns formed by them – mainly creatures and ancient gods. There were dogs and bears, there were archers, and there were dragons, serpents and scorpions. Twelve of these constellations formed a wide path through which the Sun, Moon and planets appeared to move during the year. These were the constellations of the zodiac, and they held special significance. The stars were not visible in the daytime, but it was easy to calculate which sign of the zodiac 'contained' the Sun at any given time. It was also possible to calculate in which part of the zodiac the Moon and each of the planets lay. Thus came the beginnings of astrology and the belief that the positions of the heavenly bodies in the sky affected our life on Earth. Chinese records relating to the study of the stars go back to the third millennium BC, and many of the dates in their chronology can be identified from their records of eclipses. Thus the Shang Dynasty, dating from about 1760 BC to 1122 BC, is well recorded astronomically and it contains much data from these early centuries.

Early Theories about Earth

It was only a question of time before the first theories were put forward about the origins of the world. The earliest ancient civilizations thought the Earth to be flat. They thought it was natural for everything to be drawn to the ground. There was no need to explain gravity; it was a familiar phenomenon. If voyagers travelled far enough in any direction they would find that the Earth had a limit, a place in the ocean where any ship would fall off the edge of the world or perhaps a place beyond the mountains where there was a sudden end to the land. To explain the night sky, a great sphere of clear crystal was envisaged that carried all the stars and the planets on their diurnal journey. The Sun and the Moon had their own spheres and their own independent motions. The extent of the Earth was not known, but it was obvious to all that it was the centre of the universe; it seemed larger than the Moon or the Sun and far larger than any other celestial object.

Calendars and the Gods

All the early civilizations recognized the need for a calendar. The Aztecs had two calendars, one with a solar year of 365 days and one with a sacred year of 260 days. The latter consisted of 13 sequences of 20 days, each of the sequences being ascribed to one of 20 deities.

Arithmetic shows that, after 52 years of the 365-day calendar, the 260-day calendar had revolved through 73 cycles and the two calendars had come into phase with each other again. The Aztecs studied the Moon's motion, and they were able to predict the occurrence of an eclipse. The Mayans had the same calendar as the Aztecs with the same solar and secular year. Both civilizations built temples and pyramids to the gods of the Sun and the Moon. The Incas of Peru had a similar religion to the Mayans. The importance of the Sun and Moon was recognized. Temples to the Sun were decorated with gold, and temples to the Moon were decorated with lavish silver furnishings.

Every ancient civilization built temples to their gods. In most cases it was the duty of the priests to look after the calendar, and they earned their keep by predicting the seasons and telling farmers when to plant seed and reap the harvest. The priests needed to observe the objects in the heavens and to record unusual happenings in the sky. They all recognized the importance of the Sun and the Moon. They knew that the stars appeared in the same positions and in the same constellations every night. Every early civilization had also noticed the 'wandering stars'. The Sun, the Moon and the wandering planets roamed across the zodiac from constellation to constellation. They had obvious functions, but what was the meaning of the multitude of other objects in the night sky? What was the meaning of the signs that the heavens were trying to convey to the Earth?

2

FROM BABYLON TO ANCIENT GREECE

Ancient civilizations began to create calendars as a way of reminding them about important annual events, such as the times to plant and then to store crops. As cultures developed, the calendars became more sophisticated. Later, observations from eclipses led philosophers and early astronomers to develop ways of measuring the distance from the Earth to the Sun, and even the size of the Earth itself. However, for many millennia the belief that the Earth was at the centre of the universe held back the advancement of our understanding of the true nature of the solar system.

In the fertile plain between the great rivers of the Tigris and the Euphrates there developed the ancient civilizations of Sumeria and Babylon. The Sumerian civilization, which was even older than the first Egyptian dynasty, dates back to about 8000 BC. The Sumerians developed a form of writing executed with a stylus on a clay tablet. The writing became known as 'cuneiform' text, with the symbols written with the stylus representing either nouns or verbs.

The Sumerian Calendar

In addition to this early form of writing, the Sumerians were also the first civilization to develop a working calendar. Like all who followed them they wanted to bring together the cycles of both the Sun and the Moon into their calendar. The Moon took 30 days, measured to the nearest day, to go through all its phases, and thus the Sumerian year consisted of 12 months of 30 days, giving a total of 360 days. The fourth month was written in cuneiform as the character for 'seed'; the 11th month, the harvest month, was written as the character for 'grain'; and it was followed by the character for 'house' or 'barn'. The Sumerian calendar thus predicted and noted the season to sow the seed, the season to harvest the crops, and when to store them in the barn for the winter.

For a while the calendar served them well, but after a few years they found that the harvest did not ripen by the 11th month because their year was too short by just over five days. So the Sumerians simply added an extra month to their calendar every six or seven years. They knew that their year was too short, but a year of 360

17

days had one great advantage. It was a number that divided exactly by 60, and because of this the complete circle of the heavens was seen as divided into 360 equal parts that we now call degrees. The base of 60 was used for both angular measure and time intervals – thus 60 seconds became one minute and 60 minutes made an hour. Eventually the angular measure of the degree was also divided into the 60 parts that we know as minutes, and the minutes themselves were divided into 60 seconds of arc. The confusion between minutes and seconds of arc, and minutes and seconds of time, still remains, however.

Babylonian Calendars

In the dynasty of the emperor Hammurabi (1792–1750 BC) the capital of the Sumerian kingdom moved to Babylon and the area became known as Babylonia. Much of the ancient calendar was retained, and every lunar month began on the first appearance of the crescent Moon. The astronomers went on to divide the stars into a zodiac of six houses of unequal portions.

The Babylonians were not only great astronomers, they were also very able mathematicians. We see many examples of geometry and trigonometry appearing in their work, as well as advances in astronomy. Their calculation of pi, the ratio of the circumference of a circle to its diameter, was given as three and one eighth; it was in error by a few per cent, but it shows that they knew the importance of this ratio. One of the surviving cuneiform scripts shows that the Babylonians knew how to solve the quadratic equation. They were the first civilization to introduce a seven-day week and they named the seven days after the Sun and the Moon followed by the five known planets, to give the sequence Sun, Moon, Mercury, Venus, Mars, Jupiter and Saturn. (This naming is still very evident in languages such as French and Spanish, but in English the Norse gods have supplanted all the original names except for Sunday, Monday and Saturday.)

The Earth was often imagined by many ancient peoples as a hollow hemispherical shell floating on the world-waters – the idea of the Earth as boat-shaped is one of the oldest images of the Earth. In the Babylonian conception of the universe the Earth occupied the central place and was the accepted centre of their planetary system.

According to the Babylonian system the sky forms a hollow vault above which reside the gods. In the east (left) is a door out of which the Sun rises each day, and a similar door is in the west (right) through which it returns. In the sky there are fixed stars, meteors and moving planets. The Earth is round in the form of a large hollow mountain, which rests on water. In the east is the bright mountain of the rising Sun and in the west the dark mountain of the sunset. The sea extends over the sky and in the southwest lies the island of the blessed. Under the ground is the region of the dead, consisting of seven concentric areas.

Egyptian Calendars

At about the time the Babylonians were devising their calendars, the ancient Egyptians also made valuable contributions to the calendar. For the Egyptians, the annual flooding of the Nile was their lifeblood, and to be able to predict the flooding was of great significance to them. Their calendar evolved with only three seasons, which they called the flooding, the subsistence of the river and the harvesting. Each of these three seasons was divided into four lunar months, making a total of 12 months in all. The Egyptians instigated a calendar based on their 12 lunar months. This calendar remained in use for approximately 3000 years, but like the Babylonians the Egyptians needed to insert extra months every six or seven years to maintain its accuracy. In later years they simply added five or six days at the end of the year. This was the system they used in 440 BC when the Greek traveller and historian Herodotus (484–420 BC) gave a description of the Egyptian calendar:

> *The Egyptians by their astronomy discovered the solar year and were the first to divide it into twelve parts – and in my opinion their method of calculation is better*

than the Greek; for the Greeks, to make the seasons
work out properly, intercalate a whole month every
other year, while the Egyptians make the year consist
of twelve months of thirty days each and every year
intercalate five additional days, and so complete the
regular circle of the seasons.

Herodotus describes here a calendar that had been
in use for several millennia. The Egyptians introduced a
year with 12 months each of 30 days, following the
calendar of the Sumerians. Then they simply added an
extra five feast days to make the year up to 365 days.
They knew that this made the length of their year correct
to the nearest day, but it was still too short by a little
over a quarter of a day. They stuck faithfully to their 365-
day calendar, however, allowing their months to migrate
to different seasons as the missing leap year days accumu-
lated. Every four years their calendar lost another day
until after the incredible span of 1460 years – that is, four
cycles of 365 years – it was back to where it had started!
This cycle was known as the Cycle of Sothis. It does not
seem logical that a civilization as advanced as Egypt's
would allow this to happen when they knew the exact
nature of the error in their calendar. It made nonsense
of the dates for planting the seed and reaping the crop,
but the Cycle of Sothis was executed at least twice before
the Egyptian calendar was reformed.

Early Greek Astronomy

Herodotus also knew something of astronomy in his own country of Greece. Early Greek civilization does not have the antiquity of ancient Egypt, but it still has a lineage dating back to the eighth century BC. We are very fortunate that Hesiod, the earliest of the Greek poets, gives us a wonderful description of the skies from his poem 'Works and Days', where he describes the life of the agricultural peasant:

> When the Pleiades rise it is time to use the sickle, but the plough when they are setting; forty days they shall stay away from heaven; when Arcturus ascends from the sea and, rising in the evening, remains visible for the entire night, the grapes must be pruned; but when Orion and Sirius come in the middle of heaven and the rosy fingered Eos sees Arcturus, the grapes must be picked; when the Pleiades, the Hyades, and Orion are setting, then mind the plough; when the Pleiades, fleeing from Orion, plunge into the dark sea, storms may be expected; fifty days after the Sun's turning is the right time for man to navigate; when Orion appears, Demeter's gift has to be brought to the well-smoothed threshing floor.

It is said that the Greek astronomer Thales (c.624–546 BC) predicted an eclipse of the Sun in about 600 BC. Greek astronomy was not sufficiently advanced for him to have made a long-range prediction of an eclipse at that time, but if he had been observing the Moon closely he may well have been able to make the prediction just a few days before the event. Alternatively, he may have had access to Babylonian or Egyptian records.

Life changed only slowly in the long centuries of the ancient world, and at that time astronomy also advanced slowly. Thales was a contemporary of the mathematician and philosopher Pythagoras (c.580–c.500 BC), famous for proposing his theorem concerning the square of the hypotenuse of a right-angled triangle. Pythagoras argued that the Earth was a sphere. He must have seen and wondered at the sight of a ship falling below the horizon, but his reasoning was that of the philosopher rather than the astronomer – he thought the Earth must be spherical simply because in his opinion the sphere was the perfect shape. There was a significant advance in the middle of the next century (c.450 BC) when Oenopides (c.490–420 BC) discovered that the Sun seemed to orbit the Earth in a plane inclined at 23 degrees to the rotation of the stars about the pole. It is this angle that determines the position of the tropics and also the Arctic and Antarctic circles.

The Legacy of Alexander the Great

The fourth century BC brings us to the time of Alexander the Great (356–323 BC) and his tutor Aristotle (384–322 BC). Alexander, in his conquest of the world, founded many cities bearing his name. The most famous and successful was Alexandria in Egypt, founded in the year 332 BC. It became a meeting point for Greek and Egyptian ideas and learning, as well as becoming a great trading centre. The city named after him was Alexander's greatest contribution to culture and his most important legacy. The city grew quickly in size and status and for centuries it was the greatest centre of science and culture in the ancient world. It boasted the greatest library in the world, with scrolls and documents collected from every known civilization.

While Alexander's contribution to astronomy was that he carried the learning and philosophy of the Greek world to all the countries he conquered, there was one downside – Alexander was tutored by Aristotle, a great philosopher but a very poor scientist. Aristotle considered that to actually go out and measure something was an activity more suited to a craftsman or a slave. He asserted that gentlemen could reach their conclusions just by thinking through a problem or by arguing their case in the market place or the forum. It was Aristotle who wrote that *'It does not necessarily follow that, if the work delight you with its grace, the one who wrought it is worthy of esteem.'*

He is also credited with saying *'What are called the mechanical arts carry a social stigma and are rightly dishonoured in our cities.'* By sitting in his armchair and thinking, Aristotle had decided that it was impossible for the Earth to move. We have to conclude that Aristotle was one of the men mainly responsible for the 'snobbish' attitude of Greek philosophers towards practical science; he would look down his nose with disdain at the astronomer measuring angles in the skies. His philosophy did much to delay scientific progress, not just for centuries but for millennia, and his negative influence in fields like technology and mechanics was not fully overthrown until the time of Isaac Newton.

The Hypotheses of Aristarchus

In the Aegean Sea, very near the mainland cities of Ephesus and Miletus, lay the small Greek island of Samos. It was here in the sixth century BC that Pythagoras had developed his philosophy and mathematics. It was here, too, in the third century BC that the astronomer Aristarchus (c.310–230 BC) was born. Aristarchus later moved from Samos to Alexandria, where it would be easier for him to follow his interests in astronomy. In the third century BC every educated person knew that when there was an eclipse of the Sun the event was the result of the Moon passing across the face of the Sun. But there was another kind of eclipse, less spectacular but more common than the solar eclipse, and it happened when the Earth passed between the Sun and the Moon. When the shadow of the Earth fell upon the Moon, the latter could still be seen during the eclipse, but with much reduced brightness. On the assumption that the Earth was at the centre of the universe, it was not difficult to explain the lunar eclipse in terms of the shadow of the Earth. But Aristarchus had his own thoughts about the nature of the Earth itself, and his ideas were very advanced for his time. He had no doubt that the Earth was a sphere, a view endorsed by the curved shape of the Earth's shadow, but he also proposed the heretical idea that the sphere of the Earth was much smaller than the sphere of the Sun. He formulated a set of six hypotheses before setting out to prove his revolutionary theories. Here for the first time we have a record of an astronomer using the elements of geometry to calculate astronomical distances, in particular the distance to the Sun.

1. **The Moon receives its light from the Sun.** *This was Aristarchus' way of pointing out that the phases of the Moon are a result of the illumination it received from the Sun. It was obvious to him that the Moon was a sphere in space, with the Sun shining on it.*

2. **The Earth has a relation of both point and centre to the orbit of the Moon.** *Here Aristarchus was careful not to place the Earth at the centre of the Moon's orbit. He knew that the orbit was not quite a circle, but he also knew that the Earth somehow controlled the Moon's motion and that it lay at a key point in the Moon's orbit.*

3. **Whenever the Moon appears divided in half, the great circle between light and dark is inclined to our sight.** *The great circle was the divider between light and dark on the Moon; in modern parlance it is called the 'terminator'. Aristarchus meant that at half Moon the observer on Earth was in the same plane as the circle that divided the bright side of the Moon from the dark side.*

4. **Whenever the Moon appears divided in half, the angle between Earth and Moon seen from the Sun is 1/30th of a quadrant.** *Aristarchus attempts to measure the distance to the Sun. This is a very clumsy way to describe an angle. The Babylonians had used the system of dividing the circle into 360 degrees long before Aristarchus, but in the third century BC the Greeks did not use it. A quadrant is 90 degrees. And 1/30th of a quadrant is 1/30th of a right angle, or 3 degrees. This was the angle Aristarchus used to calculate the solar distance. His method was correct*

but the true value of the angle was very much smaller than he thought – it was only about 9 minutes of arc.

5. **The width of the Earth's shadow is that of two Moons.** *If we could stand on the Moon during a lunar eclipse, we would see part of the Sun cut away by the Earth. We would be standing in the Earth's penumbra or partial shadow. When the Earth completely covered the Sun we would be in the umbra, the complete shadow. The umbra can be thought of as a cone that varies from the width of the Earth at its base to zero at the apex. During a lunar eclipse the Moon passes through this shadow cone. It has an angle of about half a degree and a length, the 'height' of the cone, of about 869,400 miles (1.4 million km).*

What Aristarchus meant by his statement was that the width of the shadow cone, at the point where the Moon entered and passed through, was twice the diameter of the Moon itself. His estimate of two Moons was very crude – an estimate of three Moons would have been more accurate. What we must admire, however, is the ingenuity of his method and the fact that his logic was correct.

6. **The Moon subtends 1/15th part of a sign of the zodiac.** *There are 12 signs of the zodiac and each one spans 30 degrees of sky. Therefore 1/15th of a zodiac gives an angle of 2 degrees. If Aristarchus means that the angular diameter*

of the Moon is 2 degrees then he is wrong by a factor of four.
This error is just not possible for an astronomer with the
status of Aristarchus. A 1/60th part of a sign of the zodiac
would be correct. We can only conclude that '1/15th' is a
simple error of transcription.

Using these six hypotheses Aristarchus tried to
measure the distance to the Moon. The shadow of the
Earth resembled a great cone, and Aristarchus knew that
the angle of this cone was half a degree, exactly the same
as the angular diameter of the Sun. When the Moon
passed through the shadow cone he calculated that the
distance it travelled in passing through was equal to two
lunar diameters. This implied that the distance from Earth
to Moon was one-third of the length of the shadow cone.
The shadow cone was about 230 Earth radii, and he was
able to calculate from this figure that the distance to the
Moon was about 72 Earth radii. This was a good approximation, but he could not complete his calculation because
he did not have an accurate figure for the radius of the
Earth.

Aristarchus went on to estimate the distance to the
Sun. Once again his method was very ingenious. He knew
that the Moon was illuminated by the light of the Sun
and that therefore when the phase of the Moon, seen
from the Earth, was exactly half, then the angle of the

Sun–Moon–Earth triangle was exactly 90 degrees. The angle between the Sun and the Moon could easily be measured from the Earth. If the distance from Earth to Moon was known, then the triangle could be solved and the distance from Earth to Sun could be calculated.

But the experiment was a failure. We can see from hypothesis 4 that he calculated the angle between the Sun and the Moon to be 87 degrees, when in fact it was less than one-sixth of a degree. The heavily cratered lunar surface made it impossible for him to decide when the Moon was exactly at half phase, and because of this he was only able to obtain the crudest figure of 20 lunar distances for the distance to the Sun. The lunar distance was again expressed in terms of the Earth's radius but this, too, was an unknown quantity.

It would appear from this that all the efforts of Aristarchus came to nothing. He did not leave behind a measure of the Earth's radius, nor of the lunar distance or the distance to the Sun. But in spite of these failures Aristarchus is remembered as one of the greatest astronomers of all time. He was an excellent practical astronomer and he was also a great theorist. His lasting claim to fame is that he was the first to propose a world model with the Sun at the centre of the solar system and the planets orbiting around it. His distances may have been wrong, but he knew that the Earth was a revolving globe following an orbit around the Sun. Aristarchus was a thousand years ahead of his time, and there is a strong case for calling the Copernican System the Aristarchian System.

Eratosthenes and Measuring the Earth

Later in the same century a man called Eratosthenes (c.276–194 BC) arrived in Alexandria to take up his post as the new librarian. He was born in the town of Cyrene in the upper reaches of the River Nile, about 500 miles (800 km) to the south of Alexandria. Eratosthenes remembered when, as a child, he and his playmates peered down into the darkness of a deep well. It was possible, for a short time at noon on just one day of the year, to see a brilliant light at the bottom of the dark well. The light was the reflection of the Sun on the surface of the water far below. In fact, the light could be seen only at noon on midsummer's day when the Sun was directly overhead. Eratosthenes knew that on the same day of the year the Sun in Alexandria did not reach the zenith. He could not repeat his childhood observations in Alexandria, but he could measure the height of the Sun at noon and show that it was an angle of 7.5 degrees away from the vertical. He knew that the Earth was a sphere and that this angle was the difference in latitude on the surface of the Earth between Alexandria and the town of Syene.

The ratio of 7.5 degrees to the full circle was the same as the ratio of the distance between the two places to the circumference of the whole Earth. If, therefore, Eratosthenes could measure the distance between Alexandria and Syene, he could easily calculate the circumference of the Earth. He estimated the distance in units called stadia – each unit being the length of a games stadium, although we cannot be sure of the exact value. Using modern units, the stadium is thought to have been about 80 metres (263 ft), and the distance from Alexandria to Syene was 10,000 stadia. This gives a value for the circumference of the Earth of 23,846 miles (38,400 km) – a very accurate determination, although we have to ask if the length of the stadium has perhaps been calculated retrospectively from the known circumference of the Earth!

Eratosthenes' work was only a generation later than that of Aristarchus. His result would have enabled Aristarchus to calculate the distance to the Moon with tolerable accuracy. By this time Aristotle's assertion that the Earth did not move had been rejected by a few enlightened people, but it was many centuries before the negative influence of Aristotle was completely overthrown.

Earlier Attempts to Measure the Earth

Eratosthenes was the first person to measure the size of the Earth to any degree of accuracy, but it is worth mentioning that there were prior claims to this measurement. The Chaldeans of ancient Babylon had a tradition that a person could walk 30 stadia in an hour; this equates to a distance of about 1.5 miles (2.4 km) per hour. In a day, therefore, a well-organized relay team could easily cover a distance of about 37 miles (60 km). The Chaldeans claimed that if a person could walk steadily at this speed for a year then they would encompass the whole Earth. The calculation gives a figure for the circumference of the Earth of about 13,000 miles (21,000 km). In fact, this figure is only just over half the true distance, but it is of little matter for the method is very unsound scientifically and the measurement could not possibly have been carried out by marching around the Earth.

There is another much later claim. In the first century BC, the astronomer Poseidonius (c.135–51 BC) measured the circumference of the Earth. His method was very similar to that of Eratosthenes, but he made use of the stars rather than the Sun. He made observations of the star Canopus as seen from Rhodes and from Alexandria. From the elevation of the star he was able to measure the difference in latitude between the two places, and he arrived at a figure usually quoted as 240,000 stadia or 11,923 miles (19,200 km). Again, this figure was far too low, and indeed it was disputed by the Greek geographer Strabo (63 BC–AD 23). It appears from one account that when Poseidonius made his measurement on Canopus the star grazed the horizon at Rhodes. Poseidonius probably knew nothing of atmospheric refraction, so this factor would create a substantial error in his measurements. Despite the erroneous calculations often made by ancient scientists, we must nevertheless admire the ingenuity of some of the methods they used.

3

THE ALMAGEST

During the second century BC, *advances in astronomy included measuring the brightness of stars as well as further attempts to discover the distances between bodies in the solar system. Another highlight was the creation of the Julian calendar which survived, despite its faults, for 16 centuries. In the second century after the birth of Christ the work of the Greek mathematician, geographer and astronomer Ptolemy brought a new understanding about the Earth, the stars and the planets.*

At the entrance to the harbour at Alexandria on the island of Pharos stood the most famous lighthouse in antiquity, designated as one of the Seven Wonders of the Ancient World. The Lighthouse of Alexandria is said to have been more than 110 metres (361 ft) high. But Alexandria was not the only harbour to boast one of the wonders of the world. In the ancient Greek city of Rhodes, beside Mandrákion harbour, there stood a colossal statue of the Sun god Helios. The statue, the Colossus of Rhodes, was said to be 70 cubits high (about 32 metres/105 ft).

The Brightness of the Stars

In the second century BC, when the astronomer Hipparchus (190–120 BC) worked in Rhodes, the Colossus was still remembered by the elderly residents of the city. At this time Alexandria remained unchallenged in the ancient world as the greatest centre of learning. However, in Hipparchus, Rhodes had an astronomer second to none at the time. Hipparchus made valuable contributions to trigonometry. He had a sound understanding of the geometry of the sphere, and he was the first to suggest that a mesh of imaginary circles that we now call latitude and longitude were the ideal coordinates for mapping the surface of the Earth. Using the same system he worked out a spherical coordinate system for the heavens very similar to that used for the Earth.

Hipparchus' greatest achievement, however, was that by studying the heavens he built up a catalogue of the stars and accurately recorded the positions of 850 of them in the night sky. It is a measure of his observing skill that he also assigned a degree of brightness to every star on a scale of one to six. Under his system magnitude one denoted the brightest stars and magnitude six indicated the faintest stars. It was the first attempt at a classification of magnitude, and the system was still in use centuries after his time.

Measuring the Solar System

Hipparchus was not unduly influenced by earlier thinking about the stars and planets. He believed that the Earth was a spinning globe in space and that it orbited the Sun once every year. Like Aristarchus (c.310–230 BC) and others before him Hipparchus wanted to measure the scale of the solar system – the distances from the Earth to the Moon and from the Earth to the Sun. If he could find these distances then he could easily calculate the size of the Sun and the Moon from their angular diameter of half a degree or 30 minutes of arc.

He was able to calculate the distance to the Moon without even leaving the island of Rhodes. He knew from the astronomical records that a generation before his time a total eclipse of the Sun had been observed and recorded at the Hellespont, near the city of Byzantium. He discovered that this same eclipse had also been observed at Alexandria, but because the latter was 500 miles (805 km) to the south, the eclipse was not total in Egypt. Hipparchus wanted to know how much of the Sun's disc was covered by the Moon in the partial eclipse at Alexandria, and the Alexandrian astronomers supplied him with the one item of information he needed. At the time of maximum eclipse one-fifth of the Sun's disc was still visible. The Alexandrians understood lengths much better than areas and, put more precisely, what they meant was that at maximum coverage, one-fifth of the Sun's diameter was left uncovered by the Moon. This meant that the angular difference between the two observations was one-fifth of the Sun's diameter – one-fifth of 30 minutes, or six minutes of arc. Put in another way, the triangle consisting of Byzantium, Alexandria and the Moon (BAM) had an angle of six minutes at the Moon. The distance AB between Alexandria and Byzantium could be measured by calculating the difference between their latitudes. The triangle could easily be solved to give the distance to the Moon (AM or BM). Eratosthenes' (c.276–189 BC) estimate of the circumference of the Earth was available to help Hipparchus make this calculation, and he knew how to make the small correction from the arc connecting the two places to the length of the chord or straight line between them. Hipparchus arrived at the figure of 67 to 78 Earth radii for the lunar distance; this was equivalent to approximately 285,660 miles (460,000 km) – a little high but still an excellent result.

Elusive Sun Measurements

Hipparchus went on to try to determine the distance from the Earth to the Sun. He used an ingenious idea based on the lunar eclipse, involving an observation very similar to that used in the previous century by Aristarchus in relation to the Moon. His reasoning was based on the shadow cone of the Earth. He knew that the diameter of the Sun as seen from the Earth was 30 minutes of arc. He reasoned that the shadow cast by the Earth was a cone with almost, but not quite, the same angle at the apex. Hipparchus knew that the small difference between these angles was the key to determining the Sun's distance. An observer at the apex of the shadow cone would be further away from the Sun and would therefore see a smaller angular diameter than an observer on the Earth. Hipparchus expected the difference between the two angles to be a few minutes of arc. His method was sound, but he was wrong about the angular difference; it was not minutes of arc he had to measure, it was a few seconds of arc and, using ancient-world technology, it was impossible to measure angles this small.

But how could Hipparchus send an observer to the apex of the Earth's shadow to view the marvellous eclipse with the disc of the Earth just covering the Sun? How could he measure the angle he wanted? He thought up a very ingenious solution. When the Moon passed through the shadow of the Earth he calculated that it should be possible to measure the width of the shadow cone by observing the time as the Moon entered and then left the shadow. He knew the distance and the speed of the Moon's motion, so he could calculate the length of the path through the shadow and hence the distance to the apex of the shadow cone. Finally he could calculate the angle of the cone.

As with Aristarchus' work there was nothing wrong with Hipparchus' logic. It was a brilliant idea, but it suffered from the same problem as the method used by Aristarchus. If the Sun had been a mere 20 times further away than the Moon then the method could have yielded results, but the difference between the angles he sought was so small that it was lost in the inevitable errors of the measurement.

Thus the problem of the true distance of the Sun was never solved in the ancient world. In fact the first reasonable estimates of this astronomical unit do not appear until the 17th century. Hipparchus did much important work, however. One of his many successful achievements was the discovery of the precession of the equinoxes. He found that the axis of the Earth's rotation did not align with a fixed point in the sky. Over a long period of time the poles traced out a small 'circle' in the sky. The Earth behaved rather like a spinning top and its axis of rotation precessed around the circle. Hipparchus was able to calculate that it took 26,000 years for the Earth's axis to complete its cycle.

The Julian Calendar

Fifty years before the birth of Christ, Julius Caesar (100–44 BC) became emperor of Rome. He recognized that the Roman calendar needed radical reform, and he employed an astronomer called Sosigenes (fl. 46 BC) from Alexandria to create a new system, to be known as the Julian calendar. Thanks to men like Hipparchus in the century before him, Sosigenes had at his disposal all the knowledge needed to create the perfect calendar. But Sosigenes is not remembered among the great astronomers of the time, and the calendar he produced looks like a classic example of design by committee. He knew that the length of the year was a little over 365 days, so he arranged for the Julian calendar to have a leap year day every fourth year. This in itself was not a sound decision, since it was known that a leap year every four years would create an error of about three days after four centuries. Although Sosigenes knew about this discrepancy and could have adjusted for it, the Julian calendar was not reformed until 16 centuries later, by which time it was ten days in error. The Gregorian calendar, instigated by Pope Gregory XIII in 1582, made allowances for the error by decreeing that the first year of each century should only be a leap year when it is divisible exactly by 400.

Flaws in the System

The Julian calendar contained 12 months. It was not possible to make the months the same length and still add up to 365 days, so it was decided that some months should have 30 days and others 31 days. However, instead of alternate 30- and 31-day months, Sosigenes allocated the 31-day months randomly. By the time the days had been allocated for the first 11 months, February – the last month in the Julian calendar year – was left with 28 days. So Sosigenes left this short month at the end of the year with an extra day on leap years only. It is said that Caesar became jealous when he discovered his predecessor Augustus had a month with one more day in it than Caesar himself had. Caesar therefore stole a day from one of the other months, leaving the calendar in chaos and posing even more questions about the state of mind of the Roman emperors! To add to the confusion, the equinoxes and the solstices all fell on the 21st day of the month. It would have been better to have shifted the calendar so that these events took place on the first day of the month, leaving three months for each of the four seasons. Logic did not prevail, however, and the result was a catastrophic calendar design, leaving posterity forever puzzling about how many days there would be in each new month, and chanting rhymes about 'thirty days hath September, April, June and November'.

Ptolemy's Great Work

There is one other astronomer of note in the ancient world. His name was Claudius Ptolemaeus (c.85–165), better known simply as Ptolemy, and he worked at Alexandria in the second century AD. His interests extended beyond astronomy into mathematics and geography. Ptolemy's astronomical work was enshrined in his great book the *Mathematike Syntaxis* (*The Mathematical Collection*). It eventually became known as *Ho Megas Astronomos* (*The Great Astronomer*). In the ninth century Arab astronomers used the Greek superlative *Megiste* to refer to the book. When the definite article *al* was prefixed to the term, it became known as the *Almagest*, the name still used today. The *Almagest* was a great work covering many aspects of astronomy and one that was destined to be the most important reference work for astronomers for more than a thousand years.

It is sometimes difficult to appreciate the timescales involved in the ancient world. Ptolemy's work is separated from that of Hipparchus by 300 years. His work is over 400 years after Aristarchus and 700 years after Thales and Pythagoras. The *Almagest* is a work of 13 books, each of which deals with certain astronomical concepts pertaining to stars and to the solar system. In essence, it is a synthesis of all the results obtained by Greek astronomy. We know that Ptolemy drew very heavily on the earlier findings of Hipparchus for his star catalogue and other aspects of his work. He has been accused of plagiarism because the bulk of his star catalogue was undoubtedly the work of Hipparchus, but in his defence Ptolemy added some new stars to increase the number of entries from 850 to 1022.

From the outset Ptolemy assumed that the Earth was at the centre of the universe. He showed how to predict the position of the Sun in the sky with great accuracy and how to predict the position of the Moon – a far more difficult problem – with tolerable accuracy. His methods were certainly good enough to enable his successors to predict both solar and lunar eclipses from his tables. The motions of the planets were very complex on account of their retrograde motion. They passed from west to east across the night sky, but at some point in their orbit they would move backwards for a few months before going forwards again. Ptolemy did not try to find a physical reason for the retrograde motion, but he had an excellent method of simulating it so that the positions of the planets could be predicted well in advance. Everything was based on circular motion, because the circle was seen to be the perfect figure. To find the path of a planet he used a system that we call deferents and epicycles, a very ingenious system of circles within circles that predicted the orbits with great precision. It was originally devised by Apollonius of Perga (*c*.262–190 BC), a brilliant Greek mathematician who lived several centuries before Ptolemy. A deferent was a large circle centred on the Earth, and an epicycle was a small circle whose centre moved around the circumference of the deferent.

Ptolemy deemed it necessary to try to explain the mechanism that held the stars and the planets in the sky. Gravitation in the Newtonian sense was not understood, but everybody knew that all solid bodies fell towards the Earth and that something in the heavens must therefore be preventing the stars from falling. The device that had been suggested by Eudoxus and others was that of a great crystal sphere rotating daily about the Earth and carrying the stars around with it. The sphere of the stars did not carry the planets, for it was well known that they did not remain at the same point in the sky. It was therefore necessary to introduce the clumsy device of an extra sphere for every planet, and yet another two crystal spheres to explain the motion of the Sun and the Moon.

An Immovable Earth, and the Motions of the Moon and the Sun

In the first book of the *Almagest* Ptolemy described his geocentric system and gave various arguments to prove that, in its position at the centre of the universe, the Earth must be immovable. He showed that if the Earth moved, as earlier astronomers like Hipparchus had suggested, then certain phenomena should be observed. He argued, for example, that since all bodies fall to the centre of the universe, the Earth must be fixed at the centre, otherwise falling objects would not be seen to drop towards the centre of the Earth. He also showed that when a body was thrown vertically upwards it always returned to the point on the Earth from where it was launched. He claimed that if the Earth rotated once every 24 hours, as suggested by Hipparchus, this would not happen.

In his second book Ptolemy described the diurnal motion of all the objects in the skies. He calculated the time that they rose over the horizon and the time when they set below it. Book three described the motion of the Sun through the zodiac.

In books four and five he demonstrated the motion of the Moon and what he called the lunar parallax. All this was taken from the work of Hipparchus. He also went on to calculate the sizes and distances of the Sun and Moon, again following the work of Hipparchus. The figures for the Moon were fairly accurate, but the Sun was taken to be only about 20 times further away than the Moon; this was the distance calculated by Aristarchus. The system still made the Sun much larger than the Earth, but Ptolemy did not see this as a good enough reason to put the Sun at the centre of the universe.

Eclipses and Star Motions

Book six of the *Almagest* was dedicated to eclipses, both of the Sun and the Moon. The eclipses were seen as major events, and it was very important for an astronomer to be able to predict them well in advance. The Ptolemaic system was quite capable of doing this, for the motions of both the Sun and the Moon were accurately described. The most difficult problem was finding the longitude of the Moon. Although the method could predict an eclipse, it did not give the correct distance of the Moon. If the lunar distance calculation had been correct, then at some points in its orbit the Moon would be seen in the sky at up to four times the size of the Sun.

In books seven and eight Ptolemy considered the motion of the stars. He described the precession of equinoxes that had been carefully measured by Hipparchus. His work was not all plagiarism, however, and Ptolemy made his own estimate of 36 seconds of arc per century for the precession. This was poor compared with Hipparchus' estimate of 45 or 46 seconds per year, and indicates that Hipparchus was the more skilful observer. (Today's figure for the precession of the equinoxes is 50.25 seconds per year.)

Planetary Motions

In book nine Ptolemy showed how to model the motions

of the planets. It has long been known that the true orbits of the planets are ellipses with the Sun at one focus. There is, therefore, some irony in the fact that it was Apollonius – who wrote a brilliant treatise on the conic sections and who knew the properties of the ellipse, the parabola and the hyperbola better than anybody else before him – who also produced the ingenious but incorrect system of cycles and epicycles for the motions of the planets.

Ptolemy's next three books gave more details of the planets with accurate figures for their orbits and their constants of motion. On Ptolemy's system the Sun and the planets all moved in same plane, called the plane of the ecliptic, but he knew there were minor deviations from his system and his final book was dedicated to what he called motion in latitude and the apparent path of the Sun against the stars.

The Ptolemaic system was not perfect, but with so many variables to choose from it was able to give a good approximation for all the planetary orbits. Ptolemy realized that the planets were much closer to the Earth than were the fixed stars. He believed in the physical existence of crystalline spheres to which the stars and the planets were attached. The sphere of the stars was not quite the outer limits of his universe, however; he suggested that there were other spheres, ending with what he called the 'prime mover', and it was this that provided the motive power for all the other spheres in his concept of the universe. Furthermore, the universe was considered to be eternal. In the Middle Ages there was some concern about how the spheres could continue to drive the solar system without running down, but medieval thought had a simple answer to the problem. Manuscripts of the time show angels turning handles on the spheres to keep the heavens rotating.

The system of Ptolemy survived the fall of Rome and the Dark Ages. The *Almagest*, or *Great Work*, was a very apt name for a system of the world that was not bettered for 13 centuries.

Chinese Astronomy and Calendar

From earliest times the Chinese astronomers kept detailed records of sunspots, eclipses, novae and – especially – comets, whose unpredicted appearance they viewed as bad omens. Star tables were produced by the 4th century BC, with a continuous record being kept from 70 BC.

The Chinese believed that the emperor must preserve a harmonious relationship with the cosmic order by means of virtuous living and the correct performance of rituals and ceremonies. This required a reliable calendar. The present calendar emerged in the 14th century BC, having evolved in the centuries before. A year had 12 months of 29 or 30 days beginning at the new Moon, with an additional month every two or three years to reconcile the lunar year with the solar year. The calendar marked predictable astronomical events such as lunar eclipses and the positions of the planets, as well as equinoxes, solstices and agricultural seasons.

4

ASTRONOMY IN THE DARK AGES

The fall of the Roman empire ushered in the period known as the Dark Ages – a time of great change and uncertainty, with new empires created in the aftermath. Much of the original scientific knowledge from the ancient world was lost to the West, but kept alive by Islamic astronomers in the East. The period also saw new advances in astronomy and mathematics, thanks to enlightened scientists such as Omar Khayyam.

In 1900 the wreck of a Greek ship was discovered off the island of Antikythera. The vessel was dated to about 80 BC. The ship alone was a valuable archaeological find, but amongst the cargo was an object that has puzzled scientists and archaeologists for years. It consisted of a frame containing a set of brass wheels, very similar to the mechanism of a clock but a thousand years before its time, and quite unlike anything else ever discovered. The best explanation so far is that the mechanism was an astronomical device used to calculate the positions of the planets – an advanced version of an instrument known as the astrolabe, which was also used to predict eclipses and to follow the motion of the Moon. The device was an isolated find, but if we accept the antiquity of the Antikythera instrument then we must accept that Greek and Roman technology in the ancient world was far in advance of what had previously been thought.

A Divided Empire

The traditional date for the fall of the Roman empire is usually cited as 476 when Romulus Augustus (reigned 475–76), the de facto emperor of the Western Roman empire, was deposed by Odoacer (435–93). The mighty Roman empire that had lasted so long and that had eventually brought Christianity to the Western world, was finally overthrown by people the Romans called the barbarians. The centuries that followed became known as the Dark Ages. Rome became divided into two empires: the East and the West. The Western empire fared badly; it became a feudal, agrarian society, and after a few generations much of the knowledge passed down from the ancient world was never used and in many cases was lost. The Eastern empire, the Byzantine empire centred on Constantinople, fared much better. Manuscripts and other forms of knowledge from the library at Alexandria found their way to the East where many scripts were copied and translated into Arabic so that they could be studied by Islamic scholars.

For more than a century after the fall of Rome the Islamic empire seemed content with its boundaries, but after the flight of the prophet Mohammed (570–632) from Mecca in 622 the empire began to expand. The city of Alexandria survived Rome by nearly two centuries, but in 641 it finally succumbed to the Arabian invaders. The sacking and burning of the great library, the greatest store of knowledge in the world, is often quoted as the barbaric work of Islam, but centuries later it transpired that this was not quite the case. Many of the manuscripts resurfaced, sometimes as the originals in Greek or Latin, sometimes as Arabic translations. When the period of military expansion was complete, the Islamic scholars became great admirers of the civilizations of the past, and were keen to retrieve as much knowledge as they could from the ancient world. Within a hundred years the Islamic empire dominated the whole of the Middle East. The empire spread westwards across Asia Minor, into Egypt, and across the north of India as far as the border of China. It also spread to the west along the southern shores of the Mediterranean, and in 771 the Moors crossed the Straits of Gibraltar and settled in the region called Andalusia in southern Spain.

Andalusia was once part of the Roman empire. Scipio Africanus (236–183 BC) conquered the region in 210–206 BC, and it eventually became the Roman province of Baetica. This province flourished under Roman rule and was the birthplace of the emperors Trajan (c.56–117) and Hadrian (76–138) and the writers Lucan (39–65) and Seneca (c.3 BC–AD 65). Roman rule lasted there until the Vandals, closely followed by the Visigoths, overran the region in the fifth century. Thus when the Moors arrived in the eighth century Andalusia already had an impressive Roman history – but it was not a history well known to the local people; it had all happened long before living memory.

Arabian and Persian Astronomers

Very little is known about the state of astronomy in the centuries immediately after the fall of Rome, but by the ninth century we see the first appearance of the fruits of Arabian knowledge. The earliest-known astrolabes, for example, appear in Islam, and an example from Damascus still survives from about 830. One of the astrolabe's uses was to determine the elevation of celestial objects above the horizon. It comprised two or more flat metal discs with calibrated scales, attached so that both, or all, the discs could rotate independently. For early navigators and astronomers it served as a star chart, a compass, a clock and a calendar. It survived for centuries and it had no equal as a navigational device until the introduction of the sextant in the 18th century. The Danjon astrolabe is what is known as a portable solstitial armillary, modified for observations of the stars. The instrument is suspended by a small hook or eye, and it consisted initially of a single ring that hung in a vertical plane. Pivoted at the centre of the ring was a rod equal in length to the ring diameter, carrying sights at either end. It could be aligned on a star or a planet, with an angular scale inscribed on the armillary ring to show the object's altitude.

Many important advances in the field of mathematics were made in the Islamic world. One of the simplest, but most significant, ideas was the introduction of a symbol for the number zero, making possible the Arabic system of numerals with a base of ten. It is the system used almost exclusively today. To appreciate what a great step forward this represents we need only look at the problems of multiplication and division thrown up by the Roman numeral system that was widely used in the West at that time. Arabian mathematics went much further than just devising the numerical system we use today. They also introduced algebra, the system of mathematics where unknown quantities are represented by symbols. Equations could be manipulated algebraically to simplify many mathematical calculations.

We know the names of many of the early Islamic astronomers, and we even know a little about their work, but it is not until the ninth century that we can identify individual astronomers and their contributions to knowledge. Albategnius (*c.*850–929) was one of the earliest of the Islam school. He was an Arabian nobleman born in Batan in Mesopotamia, and he became the leading astronomer and mathematician of his time. He drew up improved tables for the motion of the Sun and the Moon. He knew that the Earth's orbit around the Sun was an ellipse, and he measured its eccentricity. He also measured the inclination of Earth's equator to its orbital plane and he made an estimate of the number of days in the year. Albategnius made his observations over a period of 40 years from his observatory in Rakku, and they were summarized in his book *Movements of the Stars*, published in Europe long afterwards in 1537. His estimate of the year was so accurate that it was used by European astronomers in the 16th century for the Gregorian reform of the Julian calendar.

Albategnius was followed two generations later by the astronomer Al Sufi Abd al-Rahman. He was born in 903 and lived for 83 years. He was another Persian nobleman, also known by the Latinized name Azophi. He published a work called the *Book of the Fixed Stars*, which first appeared in about 964. It included a catalogue of

1018 stars, giving their approximate positions, magnitudes and colours. It obviously owed much to Ptolemy's *Almagest*, but it also contained many Arabic star names that are still in use today, although sometimes in a corrupted form. An interesting entry is the earliest-known reference to the Andromeda Galaxy, our closest galaxy. It was also at about this time that a Persian mathematician and astronomer called Abul Wafa (940–97) studied and described geometrical constructions using only a straight edge and a fixed compass. He later dubbed his instrument a 'rusty compass' because it never changed its radius. Wafa also pioneered the use of the trigonometrical functions; he compiled tables of sines and tangents at intervals of 15 arc-minutes, and he seems to be the first to have used the secant and cosecant functions. Much of his work was carried out as part of an investigation into the complex orbit of the Moon.

Astrologer al-Birini

In the early centuries it is always difficult, and usually impossible, to differentiate between astronomers and astrologers because there was only a fine boundary between them. Astrologers were keen to know the positions of the planets as accurately as possible in order to help them formulate their predictions. The astrologer al-Birini flourished at the end of the first millennium. He was born just before sunrise on 4 September 973 at the city of Kath on the River Oxus, in what is now Uzbekistan. He was interested in all branches of learning and was fluent in six languages. In 998 he moved to Gurgan on the Caspian Sea where he completed his first major work called *The Chronology of Ancient Nations*. He also wrote a *Book of Instruction in the Elements of the Art of Astrology*, dedicated to a high-ranking woman called Rayhanah. It begins with sections on geometry and arithmetic, then describes the astronomy of Ptolemy and also includes a detailed section on the use of the astrolabe. The remainder of the book concentrates on astrology and shows how many of the world's events and disasters were forecast in the stars.

A New Work on Optics

In about 1000 the Arab mathematician and physicist Alhazen (*c*.945–*c*.1040) wrote the first important book on optics since the time of Ptolemy. He rejected the older notion that light was emitted by the eye in favour of the view accepted today – that light is emitted from a source like the Sun, or reflected from an object, and then gathered into the eye. His *Treasury of Optics* was written at the end of the first millennium, but it was not published in Latin until 1572, over 500 years later, giving another example of how long some of the Arabic texts survived and how they came to be known in Europe. He described lenses, plane and curved mirrors and colours. Alhazen was also an engineer. He made an expedition to southern Egypt, sponsored by the caliph al-Hakim (985–1021), where his mission was to study possible ways of controlling the waters of the Nile. He soon realized that what he had undertaken was impossible and that the Nile could not be easily tamed. He also realized that when the bad news reached the caliph he would be executed for having failed in his task. Alhazen feigned madness upon his return. He kept up the pretence for many years until the death of the caliph in 1021.

We must also give a mention to the astronomer Arzachel (1028–87), who lived in the Andalusian region of Spain. He was the foremost 'Spanish-Arab' astronomer of his time. He carried out a series of observations at Toledo, and he presented his work in the *Toledan Tables*. He corrected geographical data from the time of Ptolemy, and in the 12th century his tables were translated into Latin. Arzachel was the first to prove conclusively that there was a small precession of the Earth's orbit relative to the stars. He measured a precession of 12.04 seconds of arc per year – a brilliant and accurate piece of observation and remarkably close to the modern accepted value of 11.8 seconds. In a later chapter we shall look at the much smaller precession of the orbit of Mercury and its importance to astronomy. Arzachel invented a novel form of flat astrolabe, known as a *safihah*, details of which were published in Latin, Hebrew and several European languages. His work was well known to Copernicus who, in his *De Revolutionibus Orbium Coelestium*, quotes Arzachel and Albategnius and acknowledges his debt to their work.

Omar Khayyam, Astronomer-poet

We now come to the best-known and the most-honoured of the Persian astronomers. He was the astronomer-poet Omar Khayyam, who lived from 1044–1122. The name *khayyam* means 'tentmaker' in Arabic, and there is some evidence that his father was indeed a tentmaker and that he himself practised this trade for a short time.

A Persian nobleman called Nizam ul Mulk was educated at the same school as Omar Khayyam, in Nishapur, the provincial capital of Khurasan. Nizam described his first meeting with Omar Khayyam:

When I first came there I found two other pupils of mine own age newly arrived, Hakim Omar Khayyam, and the ill fated Ben Sabbah. Both were endowed with sharpness of wit and the highest natural powers; and we three formed a close friendship together. When the Imam rose from his lectures, they used to join me, and we repeated to each other the lessons he had heard.

*Now Omar was a native of Nishapur, while Hasan
Ben Sabbah's father was one Ali, a man of austere
life and practice, but heretical in his creed and doctrine.*

Omar Khayyam attended other institutions of
learning, including those at Bukhara, Balkh, Samarkand
and Isphahan, but he lived in Nishapur and Samarkand
in Central Asia for most of his life. On the accession of
Din Malik Shah (1055–92) as sultan of Jalal, Omar
Khayyam was appointed court astronomer with an observatory in Esfahan. Other leading astronomers were brought
to the court, and for about 18 years Omar Khayyam super-
vised his team of astronomers to produce work of very
high quality. During this time Khayyam was responsible
for compiling astronomical tables, and he contributed to
a calendar reform in 1079. He calculated the length of
the year as 365.24219858156 days – a grossly over-
accurate figure quoted to a precision not even achievable
today. It is correct as far as the fifth decimal place, but it
is almost certainly built upon the work of Albategnius a
century before him, who in turn had access to the work
and observations of Ptolemy and the Alexandrian scholars.

Omar Khayyam is one of the select group of
astronomers who also made original contributions to the
advance of mathematics. A good example is his work on
algebra, which became known throughout Europe in the

Middle Ages. His skill as a mathematician was legendary in his time. In his book on algebra he classified many algebraic equations based on their complexity. When he came to study the cubic equation he identified no less than 13 different forms. He went on to discover a geometrical method to solve cubic equations by finding the intersection of a parabola with a circle. He studied probability, including what we now call binomial probability, and he produced figures for what we know as Pascal's triangle. He questioned whether or not a ratio should be regarded as a number. To put his work into perspective it must be said that the Romans also had the means to solve the cubic equation and they, too, used a geometric method, so Khayyam's method was probably a derivation from an earlier method. It is also well known that in the third century BC the Alexandrian mathematician Apollonius wrote a treatise on the conic sections. This was also known to the Arabians, but very few could master the ancient texts and Omar Khayyam's contribution is seen as a new development to an old problem. He extended Euclid's work by giving a new definition of ratios and showed how to handle the multiplication of ratios. He also contributed to the theory of parallel lines.

The *Rubaiyat*

Omar Khayyam not only made original contributions to

science but also to literature. In fact he is better known as a poet than as an astronomer, and he is certainly the best-known Arabian poet in the Christian world. His fame is due to the Englishman Edward Fitzgerald (1809–83) who translated into English the *Rubaiyat* of Omar Khayyam, a collection of 100 short, four-line poems, and then published them in 1859. The English version of the *Rubaiyat* has gone to several editions. This is in spite of the fact that a lot is lost by the translation into English. It has to be said that Edward Fitzgerald took a few liberties in his translation and to help with the marketing, and he wrote the first stanza entirely on his own!

> *Wake, For the Sun, who scattered into flight*
> *The Stars before him from the field of night,*
> *Drives Night along with them from Heav'n, and strikes*
> *The Sultan's Turret with a Shaft of Light.*

The *Rubaiyat* contains very little astronomy, and when it does it is only in support of the philosophy:

> LXXII
> *And that inverted bowl they call the sky,*
> *Whereunder crawling cooped we live and die*
> *Lift not your hands to it for help – for it*
> *As impotently moves as you or I.*

His best-known quatrain came when he was pondering the past and the future of the universe. He decided that it was impossible to change the past. He was able to express his thoughts far better than most:

LXXXI

The moving finger writes; and, having writ,
Moves on: nor all your piety and wit
Shall lure it back to cancel half a line.
Nor all your tears wash out a line of it.

Earlier in the chapter we had a description of Omar Khayyam as a pupil by one of his contemporaries. We now have an account of the mature Omar Khayyam, the teacher, as described by one of his pupils, Khwajah Nizami of Samarkand, who relates the story:

I often used to hold conversations with my teacher, Omar Khayyam, in a garden; and one day he said to me, 'My tomb shall be in a spot where the north wind may scatter roses over it.' I wondered at the words he spake, but I knew that his were no idle words. Years after, when I chanced to revisit Nishapur, I went to his final resting place and lo! It was just outside a garden, and trees laden with fruit stretched their boughs over the garden wall, and dropped their

*flowers upon his tomb, so that the stone was hidden
under them.*

Omar Khayyam's ten books and 30 monographs
have survived. These include books on mathematics,
algebra, geometry, physics and metaphysics.

The Fall of the Moors

For centuries, the Spanish had been eager to expel the
Moors from southern Spain. In the 11th century the fabled
warrior El Cid (*c.*1040–99) fought to drive out the Moors.
In this endeavour he had the backing of the pope, who
wished to convert the Arabs to Christianity. El Cid was
considered the perfect Christian knight: chivalrous, gentle
and magnanimous in his conquests. But nothing could
be further from the truth; he terrorized the Arabs with
his night raids. He and his men raped innocent women,
pillaged and plundered the houses and mosques and gave
no quarter. In 1135 the Muslim city of Toledo fell to the
Spanish. Rumours spread about new finds in Toledo, and
inquisitive travellers came to see what they could plunder.
It became obvious to the educated that a great centre of
culture and civilization existed there. It was also obvious
to the unbiased observer that it was the Europeans, not
the Moors, who were the barbarians in southern Spain.

In England, not long after the Norman conquest, a

monk called Adelard of Bath (*c*.1080–*c*.1152) heard a rumour that rare manuscripts had been discovered in a part of Spain. Copyists and translators from all over Europe were soon on their way to Spain to try to gain access to the knowledge. Adelard was lucky. He found his way to Toledo and there, to his joy and amazement, he discovered a wonderful library where he was able to procure rare documents for his own use. Others heard of Adelard's success and followed him across the Pyrenees and into Spain.

In 711 Muslim Arabs under the leadership of Tariq ibn Ziyad (died 720) crossed the Strait of Gibraltar from Tangier and invaded southern Spain, ending the Visigothic rule there. Henceforth Andalusia's history was closely linked with that of Morocco and the North African coast until the end of the 15th century.

It is fascinating to ask how long it took after the fall of Rome for knowledge to grow and surpass the point it had reached in the ancient world. Most historians would say that it was not until the Renaissance that mankind could claim to have gained knowledge that the ancients had not discovered. As regards astronomy, the setback was more than a thousand years. It was not until after the time of Copernicus (1473–1543) that knowledge of astronomy advanced beyond that of the ancient world.

5

THE COPERNICAN REVOLUTION

In the 16th century the astronomer Nicolaus Copernicus shook the world with his heretical assertion that it was the Sun, not the Earth, that lay at the centre of the universe. Such was the expected weight of opinion against this theory that Copernicus' views were only published after his death.

Ptolemy's *Almagest* was translated into Arabic in the ninth century, but a Spanish version of his work did not appear until the 12th century. It was translated into Latin in the time of Frederick II of Denmark (reigned 1559–88) and thus became available to the majority of European scholars. In the early Middle Ages there was still much interest in astronomy, although there were few active observers. In the monasteries the primary task of the monks was to copy the gospels into Latin and other languages. They occasionally came across scientific manuscripts, however, and sometimes these were also copied. Towards the end of the first millennium we sometimes find chronicles of historical events. The *Anglo-Saxon Chronicle* is a good example: it contains copious references to astronomical events, and when an eclipse or a comet is mentioned it is usually possible to put an exact date to the sighting.

> Thus in the year 540: *'The Sun darkened on June 20th, and the stars showed fully nearly an hour past nine in the morning.'* In 678: *'there appeared the star called a comet, in August; and it shone for three months each morning like a beam of the Sun.'* The chronicle records that in 1066: *'it happened that all through England such a sign as the heavens was seen as no man had seen before. Some men said it was the star "Comet", that some men called the long-haired star. It appeared first on the eve of Letania maior, April 24th, and so shone all seven nights.'*

Visions in the Sky

The 'long-haired star' was literally woven into the fabric of English history when it appeared prominently on the Bayeux Tapestry. Six hundred years later Edmond Halley identified the object as a comet that reappeared every 76 years. A new star appeared a few years before the comet. It was first seen on 4 July 1054, and it was so bright that for several months it was visible in broad daylight. Chinese astronomers recorded the event, but there is no mention of it in European records. Nine hundred years later it became identified with a remnant of a supernova (an exploding star) in the Crab Nebula, and for a time it became the most intriguing object in the whole of the night sky.

Comets and exploding stars are rare events, but changes to the face of Moon are so uncommon that they are virtually unknown. One summer evening in 1178 five English monks were relaxing and staring at the night sky. There was a new Moon, and as they gazed at it they noticed something very strange. A great explosion appeared to be taking place on the Moon before their eyes. The monks knew that the face of the Moon never changed, and so they realized they had witnessed a very unusual event. They did not understand what had happened, but they felt that the event must be a message of some kind from the heavens. They decided to report their findings to a higher authority, and so they made their way to Canterbury where they gained an audience with the archbishop. They swore the truth of their story under oath, and we know a little about the event because it was recorded by the chronicler Gervase of Canterbury (c.1141–c.1210):

There was a bright new Moon, and as usual in that phase its horns were tilted towards the east. Suddenly the upper horn split in two. From the mid point of the division, a flaming torch sprung up, spewing out fire, hot coals and sparks.

The Moon is covered with craters – the result of being regularly struck by objects from space in its long history. It is probable that the monks actually witnessed a large object such as a meteorite striking the Moon. Recent research suggests that the crater named after Giordano Bruno may be the result of the impact witnessed in 1178. (The Earth has suffered similar attacks in the past, but over time the weather has worn all but the largest craters away so that they are far less obvious.)

A Treatise on the Astrolabe

The Middle Ages are dotted with astronomical observations, but there are no radical new ideas about the structure of the universe recorded in this period. Astrologers were common, but few could also be called astronomers. The anecdotes can still be of interest, however. Geoffrey Chaucer (*c.*1343–1400), better known as the founder of English literature, wrote a treatise on the astrolabe, the oldest astronomical instrument. Chaucer had a son, Lewis, who it seems was more interested in mastering numbers than following in his father's footsteps and mastering words. Chaucer decided to write a treatise on the astrolabe for the benefit of his son. For one who did not have any formal education in mathematics and the sciences, Chaucer's mastery of the instrument must be admired:

> *Litel Lowis my sone, I have perceived wel by certeyne evidences thyn abilite to lerne sciencez touchinge noumbres and proporcions; and as well considere I thy bisy preyers in special to lerne the Tretis of the Astrolabie. Than, for as mechel as a philosophre seith, 'he wrappeth him in his frend, that condesendeth to the rightful preyers of his frend', ther-for have I given thee a suffisaunt Asrolabie as for oure orizonte, compowded after the latitude of Oxenford; op-on which, by mediacion of this litel treatis, I purpose to teche thee a certain nombre of conclusions apertening to the same instrument.*

A Long and Varied Education

In Chaucer's time, Britain was seen as backward and uncivilized compared with places like Italy and Spain. So when the first glimmerings of a scientific revolution appeared, it was in the cultured cities of Italy. However, it was in a most unlikely eastern European town that the astronomical knowledge of 13 centuries first came into question. Nicolaus Copernicus, the son of a well-to-do merchant, was born on 19 February 1473, in Torun, a city on the River Vistula in north-central Poland. Nicolaus was the youngest of four children. After his father's death, around 1484, Nicolaus' uncle, Lucas Watzenrode (1447–1512), took him and his three siblings under his protection. Watzenrode, who later became bishop of the Chapter of Warmia, provided for young Nicolaus' education and helped him to make his future career as a church canon.

For two years from 1491, Copernicus studied liberal arts, including a smattering of astrology, at the University of Cracow. He left the university before completing his degree, but in 1493 he resumed his studies in Italy at the University of Bologna. He stayed in Bologna for four years, studying law. For a while he lived in the same house as the principal astronomer at the university, Domenico Maria de Novara (1473–1543), who held the post of official astrologer for the city. It seems certain that Novara introduced Copernicus to Ptolemy's work – not to the original writings but to one of the later versions containing certain corrections and critical expansions of the models of the planetary orbits. These corrections were of great interest to Copernicus, and they may have suggested to him the ideas that led towards formulating his famous heliocentric hypothesis.

In 1501 we find Copernicus at Frombork in Poland, but soon he returned to Italy to continue his studies, this time at the University of Padua, where he had changed his direction of study from law to medicine. Copernicus' astrological experience at Bologna was actually a better training for medicine than we might imagine, for at that time there was so much faith in astrology that the stars were thought to influence parts of the body, and a good horoscope was considered a very valuable aid towards a diagnosis. In 1503 Copernicus received his doctorate, not in astrology or in medicine but in canon law, for he had changed his direction of study yet again.

When he returned to Poland his uncle arranged a sinecure for him at Cracow. Copernicus' duties were largely administrative and medical. He collected rents from church-owned lands; he secured military defences; he oversaw chapter finances; he managed the bakery, brewery and the mills; and he cared for the medical needs of his uncle and the other canons. His astronomical work took second place to his other duties, but it occupied all of his spare time.

Formulating a Heliocentric Hypothesis

By 1514, at the age of 41, Copernicus was regarded as a competent astronomer. In that year he was invited to offer his opinion at the church's Fifth Lateran Council on the problem of the reform of the calendar. The civil calendar then in use was the one produced 15 centuries earlier by Sosigenes under Julius Caesar. Over the centuries since it was instigated it had fallen about ten days out of alignment with the stars, and the church was concerned that it cast doubts over the true dates of crucial feast days – Easter in particular. It is unlikely that Copernicus ever offered any views on how to reform the calendar, however, because there is no record that he ever attended any of the council's sessions. In time Copernicus went to live again at Frombork. He took up residence in a tower of the cathedral house, a high observatory from where he was free to continue his astronomical studies. His revolutionary ideas had probably been forming in his mind for many years, but it was at Frombork that we first learn about them.

Copernicus is remembered for his great work *De Revolutionibus Orbium Coelestium* (*Concerning the Revolutions of Celestial Spheres*), in which he argued the case for the heliocentric universe. His idea of the Sun-centred universe was first mooted, however, in a smaller volume called the *Hypothesibus Motuum Coelestium a se Constituis Commentariolu*s (usually simply known as *The Commentary*). It was only about 20 pages in length and was never published, but in this early document are the main reasons behind Copernicus' thinking. He specified seven basic assumptions, most of which were heretical in his time:

1. **The celestial circles or spheres do not have a common centre.** *Ptolemy had introduced the idea of an 'offset' circle to help produce more accurate motions of the planets. On the Ptolemaic system they all orbited about different centres.*

2. **The centre of the Earth is not the centre of the universe, but only the centre of gravity and of the lunar orbit**. *This is the heretical assumption. But notice that the Earth has not surrendered everything to the Sun. Copernicus had no knowledge of gravity outside the Earth. He had no reason to believe that the other planets had gravity, so he accepted that everything in the universe was drawn to the centre of the Earth. It was the centre of gravity of his world model.*

3. **All the spheres revolve around the Sun, so that the centre of the world is near the Sun.** *Copernicus came to this conclusion when he discovered how much simpler the heliocentric system was than the geocentric system. His other great step forward was to recognize that the Sun must be much larger than the Earth and the other planets.*

4. **The distance to the Sun is insignificant when compared with the height of the firmament.** *Copernicus knew that the distance to the stars was much*

greater than the distance to the Sun. It was still several centuries before stellar distances could be measured.

5. **The motions appearing in the firmament are not its own motions, but those of the Earth. The Earth performs a daily rotation about its fixed poles while the firmament remains immobile as the highest heaven.** *A crucial point very well described by Copernicus.*

6. **The motions of the Sun are not its own motions, they are the motions of the Earth and out sphere with which we revolve around the Sun just as any other planet does.** *The Earth moves around the Sun as well as spinning on its axis. The motion of the Sun as seen from the Earth is the result of both factors. The Sun was stationary on Copernicus' model.*

7. **What appears to us as retrograde and forward motion of the planets is not their own, but that of the Earth. The Earth's motion alone is sufficient explanation for many different phenomena in the heavens.** *Copernicus showed, using diagrams, how the motion of the planets was sometimes retrograde or backwards. The retrograde motion was easy to explain on the heliocentric system.*

There is much repetition in these postulates, but they are consistent with each other and they indicate how Copernicus arrived at his conclusions. The seventh postulate concerns the retrograde motion of the planets; he knew that this effect could be explained very simply if the planets were assumed to orbit the Sun. He also knew that the planets Mercury and Venus were never far from the Sun, and it was obvious to him that they could not orbit around the Earth as Ptolemy had suggested. He also realized that the whole world system would be greatly simplified if the Sun were considered to be at the centre of the universe. He observed that Mars was almost as bright as Jupiter when it was near the Earth on the same side of the Sun, but when it reached the far side of the Sun it was very faint and obviously very much further away. He was very scientific in his findings and he concluded from observations of planets such as Mars and Mercury that the Ptolemaic system did not give accurate distances from the Earth to the planets.

An Orderly System

Copernicus postulated that if the Sun was assumed to be at rest and the Earth and the other planets were assumed to be in motion around it, then the remaining planets fell into an orderly relationship whereby their sidereal periods increase from the Sun in a relationship to their distance from the Sun. He calculated, very accurately, the periods for the planets to be: Mercury – 88 days; Venus – 225 days; Earth – 1 year; Mars – 1.9 years; Jupiter – 12 years; Saturn – 30 years. This theory resolved the disagreement about the ordering of the planets, but it also raised new problems. To accept the theory's premises it was necessary to reject Aristotle's natural philosophy and develop a new theory to explain why heavy bodies fall to a moving Earth. It was also necessary to explain how a body like the Earth, filled with floods, pestilence and wars, could be part of a perfect and imperishable heaven. Copernicus was working with many observations that he had inherited from antiquity and whose reliability he could not verify. In constructing a theory for the precession of the equinoxes, for example, he was trying to build a model based upon very small long-term effects, and his theory for Mercury was left with some incoherencies. Any of these considerations could account for Copernicus' delay in publishing his work, but he gave another reason in the preface to the book. He claimed that he had chosen to withhold publication, not merely for the nine years recommended by the Roman poet Horace whom he greatly admired, but in fact for 36 years, because of what he knew to be the book's heretical standpoint concerning a heliocentric universe.

Copernicus spent many years perfecting his ideas and writing *De Revolutionibus Orbium Coelestium*. He was certainly working on it by 1515, and as his ideas developed further his manuscript ran to 200 pages. He could not resist telling his ideas to some of his close friends, and during his lifetime news of the Copernican system filtered through to other astronomers. The final version of the book contained a section on the solar system followed by a star catalogue. As with his predecessors from the ancient world, he included a chapter on the precession of the equinoxes and a section on the motion of the Moon. He then devoted two sections to the motion of the planets.

Copernicus' Vision

Although Nicolaus Copernicus put the Sun, and not the Earth, at the centre of the universe, some minor details on his model were still wrong. He maintained, for example, that the orbits of the planets were perfect circles around the Sun when in fact they are ellipses. We know he was not the first to suggest the idea of the Earth orbiting the Sun, and we know he had access to the work done by Aristarchus long before him. He himself knew that he was not the first astronomer to propose a heliocentric universe either. But Copernicus was a great thinker and a man of bravery and vision. He knew that his theory would create a revolution in astronomy but he also wanted the world to know that he was right and that his system was the truth. In this respect he succeeded in his ambitions, and he takes his place amongst the greatest of the world's astronomers.

Great Minds Thinking Alike

When Copernicus was in his sixties, a 25-year-old professor of mathematics at Wittenburg called Rhaeticus heard about his heliocentric theory. In 1539 he sought out Copernicus, and the two discovered they were in complete agreement about the theory. They enjoyed each other's company so much that Rhaeticus stayed on for two years. During this time he pressed Copernicus to publish his work. But Copernicus knew there would be a great outcry when his heretical ideas were known, and he intended to delay publication until after his death. The book did indeed remain unpublished until 1543, after Copernicus had died. In the meantime Rhaeticus published a volume called *Narratio Prima* outlining the Copernican theory. Rhaeticus' work preceded *De Revolutionibus* by several years, but he gave full credit for the idea of the Sun-centred universe to Copernicus. There is a tradition that Copernicus held his own book in his hands as he lay on his deathbed – if true, it epitomizes the problems faced in those times by men with heretical new ideas about the universe.

6

CHASING THE PATHS
OF THE PLANETS

*In the late 16th and early 17th centuries two astronomers strove
to calculate the positions of the planets. Tycho Brahe spent many
years measuring the paths of the planets across the sky from his
observatory on the island of Hven, while Johannes Kepler was
able to prove from this data that the orbits of the planets were
ellipses.*

It was Christmas 1566, and in the town of Rostock in
Germany a university professor was holding a festive gath-
ering for staff and students. At some point during the
proceedings, perhaps fuelled by wine, a furious row broke
out between two of the guests, both of them Danish
noblemen. Before anyone could intervene, one of the
participants threw down the gauntlet and challenged his
antagonist to a duel. A week later tempers had still not
cooled off, and so the duel took place at an appointed

place in the nearby countryside. One of the noblemen was called Manderup Parbsjerg, and the other was called Tycho Brahe (1546–1601). When the swords were drawn and the blades were flashing in anger it was Parbsjerg who got the better of the encounter. Tycho Brahe's nose received a severe cut, his face was covered in blood and he was forced to concede. Brahe had his nose patched up with a concoction of gold, silver and wax. The doctor did a good job and, apart from a rather startling appearance that sometimes gave him an advantage, Brahe was hardly troubled by his injury. His relationship with Manderup Parbsjerg was also patched up and the two became good friends.

Astrological Shortcomings

The 'man with the golden nose' was a very successful astrologer. When he witnessed an eclipse of the Moon just two months before his duel with Parbsjerg, he forecast that the sultan of Turkey was about to die. Soon afterwards, news arrived that the sultan had indeed died. This enhanced Brahe's reputation and it also convinced him of the truth of astrology. Then news arrived that the sultan had died *before* the eclipse of the Moon and the eclipse was therefore not a valid prediction of the event. This was disturbing, and Brahe soon discovered that there were other problems with his astrological predictions. He had to admit that many of them did not seem to work. He was very confident that his methods were correct, so he concluded that the problem lay with the fact that he was unable to calculate the positions of the planets with sufficient accuracy when he was casting the horoscopes. He had two sets of tables at his disposal, one was based on the well-tried system of Ptolemy, the other was a new set of tables based on the heretical ideas of an upstart Pole called Nicolaus Copernicus, who thought that the Sun was at the centre of the universe. Brahe discovered that neither of the two systems, Ptolemaic nor Copernican, gave him accurate positions for the planets. It was bad news for his horoscopes, but he was undaunted and he made a decision to dedicate his life to calculating the positions of the planets as accurately as he could.

Brahe continued with his astronomical studies. Then he made a discovery that caused him great concern. In 1572 he witnessed a very rare event, in the constellation of Cassiopeia. A very bright new star had appeared. Brahe knew that all the world's astronomers agreed that the sphere of the stars was fixed; it had been made by God at the creation and it never changed. New stars simply did not appear. He first saw the star in November 1572 when it was brighter than Jupiter. He knew he was not in error, for it was unmistakable. For several months it was the brightest star in the sky. By December its brightness had faded to equal Jupiter. By March it had faded again, but it still ranked with the first magnitude stars. It dimmed steadily through the magnitudes, until by April 1574 it was no longer visible. What Brahe was observing was a supernova – an event so rare that it has happened only three times in our galaxy during the past 1000 years. However, Brahe was more interested in the astrological significance of the new star. It was not good news:

> *The star was at first like Venus and Jupiter, giving pleasing effects; but as it then became like Mars, there will next come a period of wars, seditions, captivity and death of princes, and destruction of cities, together with dryness and fiery meteors in the air, pestilence, and venomous snakes. Lastly, the star became like Saturn, and there will finally come a time of want, death, imprisonment and all sorts of sad things.*

On the Island of Hven

In time, Brahe decided to move from Denmark to Germany, but when King Frederick II heard about this he became alarmed to think that he might lose the services of such a wise man. So the king made Brahe a generous offer. He was to have his own private observatory. It would be built on the island of Hven, an isolated but inhabited piece of land in the straits between Denmark and Sweden.

Brahe would become landlord of Hven, and by collecting the rents from the local farmers he would have financial independence over and above his royal patronage as well as his own small kingdom. Tycho Brahe could not refuse such an offer. He began to build the most magnificent observatory the world had seen.

Brahe's observatory looked like a magical fairy-tale palace. Built in the Flemish style, it rose to 12 metres (40 ft) in height, and was surmounted by domes, spires and pinnacles sufficient to grace a cathedral. It had two semi-circular observing bays on the north and south walls. It was also a luxurious home with running water in the bedrooms. It even had a gaol – a useful facility for tenants who could not, or would not, pay their rent! The observatory became known as the Palace of Uraniborg. Brahe kept a dwarf called Jep to enhance his importance, and he also acquired two large dogs, presented to him on a royal visit by King James VI of Scotland.

Brahe was an artist as well as a scientist and craftsman, and everything he undertook or surrounded himself with was innovative and beautiful. He imported Augsburg craftsmen to construct the finest astronomical instruments. He established a printing shop to produce and bind his manuscripts in his own individual way. He induced Italian and Dutch artists and architects to design and decorate his observatory, and he invented a hydraulic pressure system to provide one of the great luxuries of the time – sanitary lavatory facilities. Uraniborg fulfilled the hopes of Brahe's king and friend, Frederick II of Denmark, that it would become the centre of astronomical study and discovery in northern Europe.

The greatest and most unusual feature of the building was the many astronomical instruments it contained. There were quadrants, sextants, armillary spheres, parallactic rules, astrolabes and clocks. On the island of Hven there was every astronomical instrument known to mankind, all made by skilled craftsmen and fashioned to the highest quality. The largest instrument was the great mural quadrant that could measure the positions of the planets to within a few minutes of arc.

Nightly Vigils in the Sky

Night after night Brahe and his assistants searched for the planets and carefully measured their positions in the night sky. Month after month the positions of more and more stars were added to a great catalogue of 777 stars – all located with greater accuracy than ever before. Night after night and year after year for 20 years the lonely vigil was kept on the island of Hven. The data were collected for what would turn out to be the last, and greatest, of the catalogues created using observations made with the naked eye, and everything was carefully recorded. But where was it all leading? What was the purpose of this great enterprise? The main object was to create a set of tables to record the positions of the stars more accurately than ever before, but it also involved plotting the positions of the planets whenever they were visible – this was in some ways a greater task, for they changed their positions nightly. But Brahe wanted to do more than simply record the nightly positions of the planets. He wanted to predict what would happen in the future as well as what had happened in the past. What he really needed more than anything was a mathematician to study the data and to formulate a new theory that could predict the planetary positions in the future. Such knowledge would be a great bonus to the science of astrology.

Then Brahe received a great setback to his ambitions. In 1588 King Frederick II died, and it became clear that his son and successor Christian IV was not prepared to patronize the eccentric astronomer any longer. Brahe was forced to continue his work funded by his own resources. He soldiered on for nearly a decade in this way, but in 1597 he finally left Hven and by 1599 he had moved to Prague. It was there that he was fortunate enough to obtain a second royal patron in the person of Rudolf II, the Holy Roman Emperor, who was prepared to pay for the publication of his works. It was also in Prague that Brahe met a much younger man called Johannes Kepler

(1571–1630). As we shall see shortly, the meeting between Brahe and Kepler was a famous and significant one, for Brahe found in Kepler someone who was capable of formulating a mathematical theory that would fit his data to the motion of the planets.

The Hypochondriac Mathematician

Our story moves to Germany, where Johannes Kepler was born in 1571 in the town of Weil de Stadt. Kepler was a small, frail man. He was near-sighted, and he was a hypochondriac. He was always plagued by fevers and stomach ailments. He was also a strange and mystical character who was very interested in astrology, and at least 800 of his horoscopes are still preserved. When he was casting horoscopes for his family he described his grandfather as *'quick tempered and obstinate'*, his grandmother was *'clever, deceitful, blazing with hatred, the queen of busybodies'*, his father Heinrich was *'criminally inclined, quarrelsome, liable to a bad end'* and his mother was *'thin, garrulous and bad-tempered'*. In later life he spent many months trying to clear his meddlesome mother of a charge of witchcraft.

In 1597 Kepler married Barbara Muller. She had been twice married before and she was twice widowed. When he first met her she *'set his heart on fire'*, but unfortunately when they tied the matrimonial knot he did not consult his horoscopes or he would have realized that the planets were in the wrong signs of the zodiac for such an event to take place. The marriage produced two children, but both died very young. Kepler became disillusioned with his wife when she told him in no uncertain terms that his precious astrology was nothing more than nonsense. He responded by accusing her of being *'fat, confused and simple-minded'*.

In spite of his personal problems and his attachment to astrology, Kepler was an excellent mathematician and he was convinced that God had created the universe with a mathematical pattern. When Kepler came to study the planets he tried to fit regular plane polygons between their orbits. But despite his efforts he was unable to find any geometric pattern that fitted.

Then, to his great delight, he found that when he modelled the problem in three dimensions instead of two, using the crystal spheres to carry the planets instead of plane circles, he could fit the five regular solids between the spheres. It seemed so perfect; the regular solids formed the framework that supported the spheres. The sphere was the perfect figure, but in terms of beauty and symmetry the regular solids ranked next. It seemed to explain why God had created the five regular solids and how they fitted into the universe. He published this finding with great enthusiasm in his *Mysterium Cosmographicum* of 1596.

*And then again it struck me, why have plane figures
among three-dimensional orbits? Behold, reader, the
invention and whole substance of this little book! In
memory of the event, I am writing down for you the
sentence in the words from that moment of conception: The Earth's orbit is the measure of all things:
circumscribe around it a dodecahedron, and the sphere
containing this will be Mars; circumscribe around Mars
a tetrahedron and the sphere containing this will be
Jupiter; circumscribe around Jupiter a cube and the
sphere containing this will be Saturn. Now inscribe
within the Earth an icosahedron, and the sphere
contained in it will be Venus; inscribe within Venus
an octahedron, and the sphere contained in it will be
Mercury. You now have the reason for the number of
planets.*

A Meeting of the Minds

Johannes Kepler first met Tycho Brahe early in the 17th
century when Brahe was living in Prague. Brahe realized
that Kepler had a good grasp of mathematics and would
be the right person to help him complete and publish the
Rudolphine Tables, named after his patron Rudolf II. On
his deathbed Brahe pleaded with Kepler to finish his work
on the planets and to publish his findings after his death.
Kepler did his utmost to oblige. He worked very hard at

studying the data, but try as he might he could not get the observations to fit the cycles and epicycles described by Ptolemy. The heretical work of Nicolaus Copernicus was of little help to him, for Copernicus had moved the Sun to the centre of the universe and thereby he had simplified the calculations, but the system devised by Copernicus was no better than that devised by Ptolemy for predicting the future positions of the planets.

In the year 1604 Kepler observed a new star in the sky. He was not the first person to observe it, however, that honour goes to a court official called Johann Brunowski, a keen amateur astronomer who told Kepler about the new star. At first Kepler did not believe the report, but when the clouds cleared from the Prague sky he could not miss it. It was 17 October and the star, in the constellation of Ophiuchus, was shining as brightly as the planet Jupiter. Kepler continued to observe the star – which became known, perhaps wrongly, as Kepler's Star – for about two years, during which time it slowly faded away. Thus it was that both Brahe and Kepler had been fortunate enough to witness a supernova. It is a surprising coincidence that these two contemporaries should both find a new supernova in the sky, when we consider that it is such a rare event.

Kepler was very moved by his astronomy and he left behind a record of his feelings:

It is true that a divine voice, which enjoins humans to study astronomy, is expressed in the world itself, not in words or syllables, but in things themselves and in the conformity of the human intellect and senses with the sequence of celestial bodies and of their dispositions. Nevertheless, there is a kind of fate, by whose invisible agency various individuals are driven to take up various arts, which makes them certain that, just as they are part of the work of creation, they, likewise also partake to some extent in divine providence. When, in my early years, I was able to taste the sweetness of philosophy, I embraced the whole of it with an overwhelming desire, and with no special interest whatever in astronomy. I certainly had enough knowledge, and I had no difficulty understanding the geometrical and astronomical topics included in the normal curriculum, aided as I was by figures, numbers and proportions ...

Kepler's Laws of Planetary Movement

Kepler used Tycho Brahe's observations when he constructed his famous laws of planetary movement. Kepler tried to fit the data for the planet Mars into an elliptical orbit instead of a circle. Eventually he found to his great joy that the data for an elliptical orbit fitted well, and it explained perfectly the errors of a few minutes of arc.

The ellipse can be defined as a slanted section through a cone. Kepler knew much about the properties of the conic sections, for he had studied the work of the Greek mathematician Apollonius of Perga.

There is some irony in the fact that Ptolemy would accept nothing but the perfect circle to describe the motions in the heavens, when all the time he had the work of Apollonius available to him. To the uninitiated, Kepler's ellipses seem just as ridiculous as his idea of the five regular polyhedrons. Why should the planets, in their orbits around the Sun, be forever following a path represented by the section through a cone? What had slanted sections of cones to do with the system of the world? But as Kepler developed his theory he discovered that this really was the case. He was absolutely right and he had made a major step forward. He formulated his three laws of planetary motion as follows:

Law 1. The orbits of the planets are ellipses with the Sun at one focus. *The focus is not at the centre of the ellipse. The ellipse has two symmetrically placed foci on its longer axis. The circle is a special case of the ellipse where both the foci coincide with the centre. The foci are so called because if a ray of light from one focus is reflected by the surface of the ellipse, then whatever its direction it will always be reflected through the second focus.*

Law 2. The radius vector sweeps out equal areas in equal times. *As the planet moves around the Sun, the line joining Sun and planet sweeps out equal areas in equal time intervals. Once the constants of the orbit are known this law can be used to predict the position of the planet at any time in the future. It is a special case of the law of conservation of angular momentum.*

Law 3. The cubes of the planets' mean distances from the Sun are proportional to the squares of their periods. *There is some uncertainty about what is meant by the 'mean distance' from the Sun, but it can be taken as the geometric mean of the maximum and minimum distances. Using this law, if we know the period of a planet (time to orbit the Sun) then we can calculate its mean distance from the Sun, and vice versa.*

Laws 1 and 2 enable the position of any planet to be predicted once the orbital plane – in other words, the period and the position of the perihelion – are known. Kepler was particularly pleased with his third law for he liked discovering numerical relationships. Kepler did not produce a popular scientific work like that of Copernicus before him or Galileo after him. Instead he edited and published the *Rudolphine Tables*. The Holy Roman Emperor Rudolf II was, like Kepler himself, more interested in casting horoscopes than in astronomy. The tables brought together the most accurate set of astronomical observations ever made (those of Tycho Brahe) and the most perfect theory for the motions of the planets (the elliptical theory of Johannes Kepler).

The tables first became available to the world at large in 1628, but scientific knowledge travelled only very slowly in the early 17th century. The first people in England to make use of the *Rudolphine Tables* were the young astronomer Jeremiah Horrocks and his friend William Crabtree, in 1639, after they had spent two years trying to make sense of earlier tables.

7

GALILEO

The Great Telescope Maker

This famous Italian astronomer believed in Copernicus' view of a heliocentric universe. Galileo's own observations of the stars and planets convinced him of it even more. But at the time such views were heretical and contrary to the account given in the Bible, and they brought Galileo into sharp conflict with the church, forcing him to recant his theory even though it was correct.

The Italian philosopher and cosmologist Giordano Bruno (1548–1600) was a follower of Copernicus. He deduced from Copernicus that the other planets in the solar system, like the Earth, were all worlds in their own right. He believed that if they orbited the Sun and were much the same size as the Earth, then it was a logical deduction that they were also inhabited. He also suggested that the stars were so far away they could, in fact, be distant suns. The Copernican system showed that our world was not unique in the universe. Bruno was not strictly a scientist nor an astronomer; he was simply someone who had reached the conclusion that the Earth may not be the only inhabited world as created by God. The Catholic Church had no doubt that Bruno was wrong in his assumption that other worlds could exist. In 1600 Giordano Bruno became a martyr to science when he was burnt at the stake for refusing to renounce his opinions.

Galileo's Early Years

Galileo Galilei (1564–1642) was also a follower of Copernicus. Although he was not burnt at the stake like Giordano Bruno, his beliefs brought him into conflict with the pope. Because of this he, too, will be remembered as someone who was persecuted for their scientific beliefs. Vincenzo Galilei, Galileo's father, was a musician who moved from Florence to Pisa in about 1563, just before Galileo was born. When Galileo was about ten his family moved back to Florence. At the age of 17 we find Galileo back again in Pisa studying medicine at the university. Galileo therefore knew both Florence and Pisa very well. His interests did not lie in medicine, however; he was much more interested in mathematics and its application to physical science. The anecdotal story of Galileo watching the pendulum swing of the chandelier in the cathedral of Pisa is well known – he recognized that the period of the swing was constant and did not vary with the amplitude (the distance from one extremity of an oscillation to the middle point), and he deduced that the pendulum could therefore be used to regulate a clock.

In 1585 Galileo's father experienced financial difficulties and as a result he could no longer support his eldest son at university. Galileo returned again from Pisa to Florence to help with the ailing family business, and he took work as a private tutor to add a little to the family finances. At this time he made a device called a hydrostatic balance, which could be used for measuring the specific gravity of bodies. It was based on the principle of Archimedes, the Greek philosopher from the third century BC and a man whom Galileo ranked far above Aristotle as a scientist. In 1589 Galileo returned again to Pisa; this time he came as professor of mathematics at the university. Although he had no formal qualifications for the job he was by this time well known, and he had frequently demonstrated his mathematical skills. He enjoyed his new profession and quickly immersed himself in the life of the university. As well as mathematics he taught astronomy, including the works of Ptolemy, and at the same time he was able to develop his interests in mechanics.

Rolling Balls and Falling Bodies

One of Galileo's experiments involved rolling balls down inclined planes and then measuring the balls' speeds as they passed various markers set up along the planes. The measurement of time was no easy matter; in the 16th century primitive watches existed, but there was no such thing as a stopwatch. In his youth, when timing the chandelier in the cathedral at Pisa, Galileo had used his own pulse to measure the time intervals. As a young man he measured small time intervals using a simple pendulum of his own – a simple device based on his observations of the one at Pisa. He was able to formulate the concepts of velocity and acceleration and to show that the speed of his rolling balls increased uniformly as they rolled down the inclined plane.

Galileo turned his thoughts to bodies in free fall. He reasoned that all bodies accelerated as they fell to Earth, but that they all fell at exactly the same rate. If, for example, a large object and a smaller, lighter object were dropped together from the same height, they would both strike the ground at much the same time. How could Galileo resist using the Leaning Tower of Pisa for his experiments? What better place could he have for testing his theories about falling bodies? He dropped a cannon ball and a musket ball from the same height from the highest level of the tower. We now know that the effects of air resistance would cause the heavier ball to reach the ground before the lighter one. So the question facing us is: did Galileo ever perform the experiment or is the story apocryphal? Viviani, Galileo's first biographer, states that he repeated the experiment many times, and this seems in keeping with the truth, for it was his nature to experiment using different masses and other refinements.

Bringing the Heavens into Focus

Galileo never married, but he had two daughters and a son by a woman called Marina Gamba. His father died in 1591, and in that year he moved from Pisa to Padua university and soon afterwards he moved from Padua to Venice.

It was during his time at Padua and Venice, in the first decade of the new century, that Galileo heard about a wonderful new instrument fashioned by a spectacle maker called Hans Lippershey (1570–1619), who had a practice at Middelburg in Holland. By peering through a tube containing lenses this instrument somehow made distant objects appear to be nearer and larger. This amazing device was the telescope. Soon Galileo had discovered how the telescope was constructed and was making his own version of the instrument, which he sold to merchants and others.

Galileo was pleased with his commercial success, but

then went on to develop his instrument even further and to use it to examine the skies. He was not the first astronomer to use the telescope for this purpose, but he had a great flair for the instrument. He made his astronomical telescopes with two convex lenses. This required a longer tube but gave better results, even though the convex lenses produced an inverted image. He was soon making new discoveries.

Galileo turned his telescope towards the Moon, marvelling at how close the telescope seemed to bring it. He saw craters and mountains and what he thought were seas. All these had been seen before but never with such brightness and detail. He looked at the planets, and on nearly every one of them he saw something new. By looking in the direction of Jupiter he could see four small spots of light near the planet. He observed Jupiter every night and discovered that the spots of light changed their positions. He had discovered the four largest moons of Jupiter, namely Io, Europa, Ganymede and Callisto. This in itself produced a problem for the geocentric traditionalists because it brought the number of bodies believed to be orbiting Earth to more than the sacred number seven. Then Galileo turned his telescope to look at Saturn. He discovered two strange companions, one on each side of the planet. What he was observing were the rings of Saturn, but the telescope was not powerful enough to

resolve them properly. Over a period of time the companions grew smaller and disappeared, only to return again
later in the year (an effect caused by the rings being seen
edge on). When he observed the planet Mars he saw that
it clearly displayed a disc. However, it was the planet
Venus that offered the greatest surprise. Through the telescope Venus showed phases, just like the Moon. This observation more than any of the others convinced Galileo that
the Copernican system was right; the phases of Venus
exactly matched the motion of the planet around the Sun.

When Galileo turned his telescope to the stars he got
another surprise. The stars still appeared as tiny spots of
light – the telescope did not seem to bring them any
nearer – but when he looked at the spaces between the
stars, more and more stars appeared. When he trained
his magic tube on the Milky Way he saw new stars
appearing in their hundreds and thousands:

*In order that you may see one or two proofs of the
inconceivable manner in which they are crowded
together, I have determined to make a case against two
star-clusters, that from them as a specimen you may
decide about the rest. As my first example I had determined to depict the entire constellation of Orion, but
I was so overwhelmed by the vast quantity of stars
and by want of time that I have deferred attempting*

*this to another occasion, for there are adjacent to, or
scattered among the old stars more than 500 new stars
within the limits of one or two degrees … As a second
example I have depicted the six stars in the constellation Taurus, called the Pleiades … near these lie more
than forty others invisible to the naked eye, no one of
which is more than half a degree off any of the afore-
said six, of these I have noticed only thirty-six in my
diagram.*

The universe obviously contained far more stars than
anybody had ever imagined. The number of visible stars
in the universe seemed to have increased a thousand-
fold or more, just by the invention of the telescope. It
seemed that it would take a hundred men a lifetime to
catalogue all of them.

Getting around the Censors

Excited by his discoveries, Galileo wrote a small book called
the *Sidereus Nuncius*, or *Starry Messenger*, in which he
described them. The book appeared in print in 1610, and
not surprisingly it came in for immediate criticism. The
first salvo was fired in 1612 by a Dominican friar called
Nicolo Lorini (fl. 1614) and the second in 1614 by another
Dominican called Tommaso Caccini (1550–1618). The affair
simmered for a few years, until Galileo was summoned

to see the pope about his unconventional thoughts. In 1616 he was instructed very clearly to desist from putting forward his view that the Sun lay at the centre of the universe. But Galileo knew that he was right. Instead of renouncing his ideas he simply gathered more evidence to support his case. The result was that Galileo's *Sidereus Nuncius* was put, along with Copernicus' *De Revolutionibus Orbium Coelestium*, on the list of prohibited books.

Undaunted by these setbacks Galileo started work on his next book. In this he had support from an old friend called Benedetto Castelli (1578–1643) who was appointed as the official mathematician to the pope and also had the approval and support of Cardinal Barberini (1568–1644). Galileo meant to call his book *Dialogue on the Tides,* but under pressure changed the title to *Dialogue Concerning the Two Chief World Systems*. The book's subject was almost exactly what its title suggests, and was a trialogue with three main characters. One of the main characters was called Fillipo Salviati. He was a real person, an old friend of Galileo's who had died in 1614 at the age of 31, and so could not be persecuted for his beliefs. In the book Salviati proposes Galileo's views, putting forward the case for a heliocentric universe. He 'uses' all the evidence that Galileo had collected from his study of Copernicus, from his own telescopic observations and from other sources. Salviati is, of course, essentially Galileo himself putting forward his own case for the system of the world – a heliocentric universe.

The second character is called Giovanni Francesco Sagredo. He was also a real person and another friend of Galileo's. He, too, had died before the book was written. In Galileo's *Dialogue* Sagredo does not hold strong opinions about the system of the world, but merely acts as a kind of mediator in the discussion between the two other parties. The third character is called Simplicio. He is the defender of the geocentric universe theory in which all the heavenly bodies revolved around the Earth. It was an unfortunate but quite deliberate choice of name by Galileo for it suggested a simpleton, and although Simplicio puts forward some very clever ideas and reasons for his beliefs, he is consistently defeated in his arguments by the better-educated and well-informed Salviati. It was but a small step to associate Simplicio with Pope Urban VIII, and Galileo must have realized that the pope would be offended by the parody presented in his character, but he still went ahead with the publication.

In *Dialogue* Salviati questions the shape of the Earth. He argues that it is a sphere just as the Sun and the Moon are spheres. Simplicio refutes the idea and quotes Aristotle as his authority:

It is vain to inquire as you inquire, as you do, what part of the globe of the Sun or Moon would do if separated from the whole, because what you inquire would be the consequence of impossibility. For, as Aristotle demonstrates, celestial bodies are invariant, impenetrable and unbreakable; hence such a case could never arise. And even if it should, and the separated part should return to the whole, it would not return thus because of being heavy or light, since Aristotle also proves that celestial bodies are neither heavy nor light.

This of course is exactly what Salviati, in the person of Galileo, wants to hear and he sets about deriding the attitude of the Greek scientists who were no more than armchair philosophers, too proud to take measurements or to seek out the truth for themselves.

On Trial for Heresy

The *Dialogue* was published in 1632. Galileo had no trouble getting his book past the Florentine censors, but when it reached Rome there was a sudden turn of the tide against him. Galileo knew he had enemies, but he thought that Barberini was on his side. Barberini, however, had become Pope Urban VIII in the 16 years since Galileo started writing his book, and now found the papal stance on the issue of what lay at the centre of the universe ridiculed. Thus, although Galileo had gone to great lengths to obtain approval before publishing his views, he nevertheless found that his opponents were determined to trump up a charge against him. In September 1632 the holy office put Galileo on trial for heresy. He knew what had happened to Giordano Bruno and that he would face the Inquisition. He was found guilty of teaching the philosophy that the Earth moved. He was forced to read out a long and humiliating recantation of his views in the hall of the convent of Santa Maria Sopra Minerva before the entire congregation of the holy office. There may be no truth in the story that as he left the Inquisition the dispirited Galileo murmured *'Eppur si muove'* (And yet it moves), but the story is entirely in keeping with his views, and in the months that followed he must have said the words many times over to himself.

Galileo's trial for heresy proved to be just as disastrous for his accusers, the Catholic Church. The church was expected to defend the version of creation as told in Genesis, and the trial of Galileo was the first occasion on which anything so profound had ever challenged the literal truth of the story told in the Bible. History shows that Galileo was not guilty of heresy, but merely seeking the truth, and many years later the Vatican offered a long-overdue apology.

Theologians had always argued and debated about the interpretation of the gospels, and this frequently led to the formation of new sects and religious orders. This was particularly true in the 16th century when the Protestants made the break with Rome. After this had taken place, astronomers in Protestant countries could build upon the work of Galileo without fear of persecution. Other evidence was waiting to undermine the church's beliefs, however. For example, when geologists first began to challenge the age of the Earth from the dating of the rocks, and later in the 19th century when Charles Darwin first published his theory of evolution.

8

JEREMIAH HORROCKS

Father of English Astronomy

The story of astronomical discovery now moves to England where, free from the religious dogma that had bedevilled earlier Italian astronomers, scientists could build on theories such as heliocentrism. At the forefront of these endeavours was the momentous and prolific work of Jeremiah Horrocks, rightly described as 'the father of English astronomy'.

The Mersey spreading and presently contracting its stream from Warrington falls into the ocean with a wide channel very convenient for trade, where opens to view Litherpole, commonly called Lirpoole, from a water extending like a pool, according to the common opinion, where is the most convenient and most frequented passage to Ireland: a town more famous for its beauty and populousness than for its antiquity;

its name occurs in no ancient writer except that of Roger of Poictou who was lord, as stated of Lancaster, built a castle here, the custody of which has now for a long time belonged to the noble and knightly family of Molineux ... This Roger held, as appears in the Domesday book, all the lands between the rivers Ribble and Mersey.

Once a Beautiful Landscape

The above description of Liverpool, found in the journal *Britannia* compiled by historian William Camden (1551–1623), may not sound very much like the sprawling English city of today. The account was written in the 1580s when the population numbered less than a thousand. Early in the following century, when Jeremiah Horrocks (1618–41) was born there, the number of residents had still not reached four figures. When Camden praised Liverpool's beauty he was not guilty of any form of flattery. By the standards of the times Liverpool was a very clean and attractive seaside town with fine beaches of golden sand. To the north there were virgin sand dunes that stretched nearly 20 miles (32 km) along the coast – much further than the eye could see. To the southwest lay the rural Cheshire peninsula of the Wirral, bounded by the River Mersey and the River Dee. The vista further south showed the snow-capped peaks of Snowdonia in Wales. To the north were the mountains of the Lake District, and on a clear day Snaefell and the mountains of the Isle of Man could be seen from higher vantage points.

Near the castle there was a small natural inlet called the Liver Pool, and it was here that the shipping was accommodated. The pool was fed by a small stream, which ran through a leafy dale known as Dale Street. There was a small bridge giving access to the area 'over the pool', and here a small Puritan community had established itself in an area called Toxteth Park.

The Puritans were tolerated in the England of the 1620s, but because of their unorthodox approach to religion they were not allowed to hold high public office. It was common, therefore, for them to put their energies into trade. Thus James Horrocks was a watchmaker and his wife Mary Aspinwall was the daughter of a watchmaker. This was a time long before the Industrial Revolution; Liverpool and Lancashire had never yet seen a bale of cotton. It is very possible that the Horrocks and Aspinwall families were not only manufacturers of watches, but also dealers who marketed watches made in Nuremberg and other European centres.

When their first son was born in 1618, James and Mary Horrocks christened him Jeremiah. The couple seemed to have a penchant for Old Testament prophets of doom, for when their second son was born three years later they called him Jonah. Both sons were expected to enter the family business, but Jeremiah showed an early interest in philosophy and other subjects beyond watch-making.

A University Education

At the age of 14, with the help of his family and the local minister, Richard Mather, Horrocks had acquired sufficient knowledge of the scriptures to gain a place at Emmanuel College, Cambridge – the most puritanical of the Oxbridge colleges. Horrocks did not go to Cambridge to study astronomy, however. At this time it was not possible to study the subject at Cambridge. There was no department of astronomy and no professor of astronomy. Indeed, there were very few who knew anything at all about the subject. One of Horrocks' friends and contemporaries, John Wallis (1616–1703), who matriculated in the same year, arrived to study mathematics. The status of mathematics was much the same as that of astronomy, and Wallis described it:

> I did thenceforth prosecute it [mathematics], (at School
> and in the University) not as a formal study, but as
> a pleasing Diversion, at spare hours; as books of Arith-
> metick, or others Mathematical fell occasionally in my
> way. For I had none to direct me, what books to read,
> or what to seek, or what methods to proceed. For

mathematics, (at that time with us) were scarce looked upon as Academicall studies, but rather mechanical; as the business of Traders, Merchants, Seamen Carpenters, Surveyors of Lands, of the like; and perhaps some Almanac-makers in London. And amongst more than Two hundred Students (at that time) in our College, I do not know of any Two (perhaps not any) who had more of Mathematics than I, (if so much) which was then but little; And but very few, in that whole University. For the study of Mathematics was at that time more cultivated in London than in the Universities.

The same educational shortcomings could be levelled at astronomy. All undergraduates, if they were not of the aristocracy, were expected to train for the church and to become country parsons. The library shelves were straining with theological publications, but there was hardly a single volume on mathematics or astronomy. And yet, by the time he left Cambridge in 1635, Horrocks had read many of the latest astronomical publications and knew exactly what he wanted to do with his life. In Horrocks' time the total complement of Emmanuel College was between 200 and 300 people. He would therefore have known all of his contemporaries. His closest friends were John Worthington and John Wallis. They both went on to become active members of the Royal Society. Amongst his other acquaintances was Ezekiel Cheever (c.1614–1708), the son of a London spinner educated at Christ's Hospital School. Cheever entered as a sizar (an undergraduate who received aid from the college for maintenance in return for performing various duties) the year after Horrocks. He left for America in 1637 and became the best-known teacher in the early history of Massachusetts. Another, even better-known, contemporary was John Harvard (1607–38), who later emigrated to the New World. When he died, he left his substantial private library and half of his estate towards the foundation of a new college, later to be called Harvard College.

Stars and Poetry

In 1635, at the age of 17, Jeremiah Horrocks returned to his native Lancashire. He had decided that what he wanted from his life was to be able to study the stars and the planets. He needed books on astronomy to achieve his aims and he also needed instruments. The most important astronomical instrument of the day was the telescope; it had been in use for about 20 years and was beginning to become far more readily available. Horrocks tells us that he purchased a 'half crown' telescope, probably at a local fair. He knew that better instruments were available, however, and in May 1638 he wrote that *'I have at last obtained a more accurate telescope.'*

We now discover that Jeremiah Horrocks was more than simply an astronomer. He was also a poet. The most prominent astronomer before this time who could also claim to be a poet was the philosopher Omar Khayyam (1044–1122). But when it came to putting his thoughts into verse Jeremiah Horrocks was the equal of his Persian predecessor. He was very thrilled with his new telescope and wrote about it thus:

Divine the hand which to Urania's power
Triumphant raised the trophy, which on man
Hath first bestowed the wondrous tube by art
Invented, and in noble daring taught
His mortal eyes to scan the furthest heavens.
Whether he seek the solar path to trace,
Or watch the nightly wanderings of the Moon
Whilst at her fullest splendour, no such guide
From Jove was ever sent, no aid like this
In brightest light such mysteries to display;
Nor longer now shall man with straining eye
In vain attempt to seize the stars. Blest with this
Thou shalt draw down the Moon from heaven, and give
Our Earth to the celestial spheres, and fix
Each orb in its own ordered place to run
Its course sublime in strict analogy.

During his time at Cambridge, Horrocks corresponded with Herbert Gellibrand (1597–1637), the professor of astronomy at Gresham College in London. Gellibrand, acting in good faith, suggested to Horrocks that he purchase a copy of a book by a Belgian astronomer called Philip Lansberg (1561–1632). Horrocks followed this advice, and he spent the next year trying to fit his observed motions of the planets to Lansberg's tables. Horrocks met with no success in this endeavour, but in 1636 he befriended a fellow amateur astronomer called William Crabtree (1610–44), who was working at Broughton near Manchester. It was Crabtree who suggested to him that he use the *Rudolphine Tables*, the work of Johannes Kepler and Tycho Brahe.

The incident illustrates how long it took for scientific works to circulate in the early 17th century. Gellibrand apparently knew nothing of the works of Kepler and Brahe, or he would certainly not have recommended Lansberg's tables to Jeremiah Horrocks. We now know that the *Rudolphine Tables* were far in advance of anything else available at that time, but the first people in England to use them were the amateur astronomers Jeremiah Horrocks and William Crabtree. The Copernican system also took a long time to circulate. The Ptolemaic system was still in use by the astrologers at this time, but Horrocks had little regard for it. He soon discovered that the

Copernican system was far superior and he put his feelings into verse:

> Why should'st thou try, O Ptolemy, to pass
> Thy narrow-bounded world for aught divine?
> Why should thy poor machine presume to claim
> A noble maker? Can a narrow space
> Call for eternal hands? Will thy mansion
> Suit great Jove? or can he from such a seat
> prepare his lightnings for the trembling Earth?
> Fair are the gods you frame forsooth! nor vain
> Would be their fears if giant hands assailed them.
> Such little world were well the infant sport
> Of Jove in darker times; such toys in truth
> His cradle might befit, nor would the work
> In after years have e're been perfected,
> When harlot smiles restrained his riper powers.

A Move to the Country

In 1639, at the age of 21, Jeremiah Horrocks left his home in Toxteth. He moved to the small village of Much Hoole about 18 miles (29 km) north of Liverpool. Nobody has been able to find the reason for Horrocks' move. Some have suggested that he was offered the post of curate at the chapel of Much Hoole, whilst others have suggested that he was employed as a tutor at Carr House in

Bretherton. It would be fascinating to discover that the reason was a romantic attachment, but there is no solid evidence to support this theory, either. All we know is that from the summer of 1639 his correspondence to his friend William Crabtree was addressed from Much Hoole and not from Toxteth.

In a survey taken in 1641 the population of Much Hoole was recorded as 235 adults. The choice of residences was limited, with the vernacular-style farmhouses being low and thatched, with smoking peat fires. The dwellings usually had pigs and chickens running in and out of the doorway. They were a far cry from the ideal residence for an educated astronomer. Apart from the church itself the only building in the village suitable for an observatory was Carr House at Bretherton, the home of the Stones family. This was situated about half a mile (about 0.8 km) away from the church. Much Hoole is the last place where we might expect to meet merchants from great trading centres like London and Amsterdam, but there is sometimes a tendency to overplay the isolation of the English village. Above the doorway at Carr House is an inscription in stone that proves the trading connections to be a fact:

Thomas Stones of London haberdasher and Andrewe
Stones of Amsterdam marchant hath builded this howse

> *of their own charges and giveth the same unto their*
> *brother John Stones: Ano Domini 1613 Lavs.*

The Stones family were evidently haberdashers. There is a strong connection between haberdashery, drapery and cloth dealing. Horrocks' contacts John Worthington and William Crabtree were both involved in the clothing industry.

A Momentous Observation

As Horrocks continued his observations of the planets he discovered to his great delight that Venus seemed to be on course for a conjunction with the Sun – in other words, the planet seemed to be following a path that would take it across the face of the Sun. This was an event so rare that it had never been recorded before. Horrocks knew that his observations might be marred by cloudy skies, so he asked both William Crabtree and his brother at Toxteth to try to make the observation as well. He asked William Crabtree to inform Henry Gellibrand in London about the event, but it appears that the request did not arrive in time.

The year was 1639, and from his calculations Horrocks did not expect the transit of Venus to take place before 3 o'clock on the afternoon of 24 November. It appeared from the tables of other astronomers, however, that it might occur somewhat sooner, and in order to avoid the possibility of disappointment, he began to observe the Sun from about midday on 23 November. As expected, he saw no sign of the image of Venus. The next day he continued his vigil until he tells us he was *'called away by business of the highest importance, which could not with propriety be neglected'*. This phrase has given rise to the possibility that he had to give a sermon just as the transit was about to begin. He tells us nothing else about this important business, but since it was Sunday it is reasonable to deduce that he needed to perform a Sabbath duty of some kind. Nevertheless, the task cannot have been too time-consuming for he was back at his telescope again in just over an hour. This would have been just enough time for Horrocks to get to the church, perform his duties and then return to his observations.

When Horrocks returned to his observations he was overjoyed to see that a dark, round spot was already fully entered upon the image of the Sun. It was without doubt the silhouette of Venus that he had been anticipating. He did not want to be accused of seeing nothing more than a sunspot, however – even though that observation alone would have put him in the company of his mentor Johannes Kepler. Thus later, Horrocks went to great lengths in his treatise to explain that Venus appeared on the Sun's disc as a perfectly circular dark spot. As well as being a perfect circle, the spot moved across the Sun much faster than a sunspot; there was no doubt that he was observing the planet Venus. He measured the size of the dark spot as accurately as he could and he drew it in the exact position it appeared on his image of the Sun. He drew two more images and recorded the times as 3.15 pm, 3.35 pm and 3.45 pm. The image moved by one diameter in the first 20-minute interval, but slightly less in the second interval. Then the Sun set over the Ribble marshes. He knew the value of accurate measurements and he wanted his observation to be as precise as it could possibly be. He was working to angles within seconds of arc. He estimated that the diameter of Venus was 1' 12" and he estimated his error was 4 or 5 seconds of arc.

An Observation Made Elsewhere

At Broughton, William Crabtree was also trying to observe the event. He had been very unfortunate with the weather, however. The skies were overcast for the greater part of the day and the Sun was not visible. Crabtree had almost given up on the task when, a little before sunset, at about 35 minutes past 3 o'clock, at the same time as Horrocks was making his observation, the Sun suddenly burst out from behind the clouds. Crabtree rushed into his house and he began to observe at once. To his great joy he saw the rare spectacle of Venus passing across the

Sun's disc. In a passage that does much to illuminate the personalities of both men Horrocks recorded the feelings of his friend:

> *Rapt in contemplation, he stood for some time motion-less, scarcely trusting his own senses, through excess of joy; for we astronomers have as it were a womanish disposition, and are overjoyed with trifles and such small matters as scarcely make an impression upon others; a susceptibility which those who will may deride with impunity, even in my own presence, and, if it gratify them, I too will join in the merriment. One thing I request: let no severe Cato be seriously offended with our follies; for, to speak poetically, what young man on Earth would not, like ourselves, fondly admire Venus in conjunction with the Sun.* Pulchritudinem divitiis conjunctam? *[beauty conjoined with wealth] What youth would not dwell with rapture upon the fair and beautiful face of a lady whose charms derive an additional grace from her fortune?*

Jeremiah Horrocks wrote up his account of the transit of Venus, making full use of his careful observations. Firstly he was able to calculate more accurate values for the orbit of Venus and secondly he had a very good estimate of the angular diameter of Venus at the planet's closest approach to the Earth. But Horrocks went much further than this. Some may say in fact that he went too far, for he used his result to try to estimate the distance from Earth to Venus, and hence calculate the scale of the solar system. To assist him in his endeavours, Horrocks had at his disposal an account of the transit of Mercury, observed a few years before him by the French astronomer and mathematician Pierre Gassendi (1592–1655).

He then formulated what we might call Horrocks' hypothesis – that every planet has the same angular diameter when it is seen from the Sun. Horrocks knew that some of the planets did not seem to fit his hypothesis, the most obvious being Mars, which was much too small. The hypothesis never reached the status of a law but it was not an unreasonable postulate. It gave a value for what Horrocks called the parallax of the Sun, a measure closely related to what we would call the astronomical unit – the mean distance from the Earth to the Sun.

Horrocks' Table

The table shown overleaf was compiled by Horrocks to enable him to compare his estimate of the solar parallax with those made by other astronomers. In this table Horrocks shows the solar parallax in terms of radius and not diameter, and they are therefore only half of the values quoted elsewhere. An extra column is included to show the distance in millions of kilometres, and the bottom line has been added to show the modern accepted values.

Astronomer	Parallax of the Sun	Distance	Astronomical Unit	Sun: Earth
	(Earth radius/ AU)	(Earth radii) Radii	(Millions of km)	(Volume ratio)
Ptolemy	2' 50"	1210	7.7	166
Albertegus	3' 0"	1146	7.3	142
Copernicus	2' 55"	1179	7.5	162
Tycho Brahe	2' 54"	1183	7.5	139
Longomonatus	2' 35"	1334	8.4	196
Lansberg	2' 13"	1551	9.86	343
Kepler	0' 59"	3470	22.1	3,469
Horrocks	0' 14"	15,000	95.4	309,333
Modern value	0' 8.5"	24,000	149	1,200,000

The Orbit of the Moon

Thus, with the transit of Venus and his estimate of the solar parallax, Jeremiah Horrocks was the first Englishman to contribute significantly to the history of astronomy. But his main contribution was to come. He wanted to find a better method of calculating the orbit of the Moon. Apart from briefly visible bodies such as comets and meteors, the Moon was the fastest-moving object in the sky seen by early astronomers. Many had tried to explain its motion but without success. It was well known that if an accurate theory of the Moon's motion could be found then it would be an invaluable aid to navigation. Ptolemy had attempted to tackle the task, and his system at least enabled astronomers to forecast an eclipse, but if his theory had been correct then the Moon would have appeared about four times larger than it actually was at some points of its orbit. Horrocks knew that as a first approximation the Earth–Moon system was similar to a sun–planet system. The Earth was at one focus of an ellipse. The ellipse was perturbed, however, by the gravity of the Sun. Horrocks proposed a system whereby the orbit of the Moon oscillated throughout the year, and he set about trying to find the constants of this oscillation. His theory of the motion of the Moon was the most advanced of his time. It was used by the first Astronomer Royal, John Flamsteed (1646–1719), and his successors at the Greenwich Observatory, and it survived for almost a century before a better theory was found.

Posthumous Fame

Unfortunately, Jeremiah Horrocks died suddenly in 1641. Many of his works were lost in the chaos that ensued during the English Civil War (1642–9). Luckily his account of the transit of Venus survived and it was published in

1662. Thereafter, many of his other surviving works also found a wider audience, and his discoveries were acknowledged and acclaimed by the Royal Society as well as by some of the world's greatest astronomers, including Isaac Newton in his third book of the *Principia*. John William Herschel was so impressed that he described Horrocks as *'the pride and boast of British astronomy'*.

Johannes Hevelius – Map Maker of the Heavens

Polish-born Johannes Hevelius (1611– 87) was one of the leading observational astronomers of the 17th century. He came from a wealthy family of brewing merchants in Danzig, where he was also a town councillor and later mayor of Danzig.

However, his main interest was in astronomy. He built an observatory on the roofs of three connecting houses, the pattern of which closely followed that devised by Tycho Brahe on Hven. As well as numerous other astronomical instruments (all described in his publication of 1673 *Machina coelestis*) his most notable achievement was the construction of a 45-metre (150 ft) telescope of his own design.

In 1647, after ten years of observation, he produced detailed maps of the Moon, published in his work *Selenographia*, including diagrams of phases and first estimates of lunar mountain heights. He also made many observations of comets published in *Prodomus Cometicus* (1665) and *Cometographia* (1668). During his life he mapped positions for 1564 stars, and these were eventually published posthumously in 1690 by his second wife, Elizabeth Margarethe, in a catalogue, *Prodomus Astronomiae*, and an atlas, *Uranographia*.

9

THE CLOCKWORK
UNIVERSE

It was Christmas Eve 1642 when Hannah Newton first felt the birth pains. Her child was not expected until January or February but the contractions became regular and consistent. The child was a boy, born on Christmas Day. Hannah was a widow. She had experienced marriage, death and birth all within the space of nine months. She chose to call the child Isaac, after his deceased father.

The Newtons lived at Woolsthorpe Manor in the village of Colsterworth in Lincolnshire. It lay just off the Roman road of Ermine Street, which was at that time a part of the Great North Road from London to Scotland. The young Isaac Newton (1642–1727) was precocious; he enjoyed reading and he enjoyed constructing things. He received the rudiments of an education, but at the age of 14, when he was old enough to help with the farm, his mother took him away from school. It was a mistake. Young Isaac was discovered reading under the hedge when he should have been tending to the sheep. Sometimes he wandered around in a dream. On one occasion he walked all the way home from Grantham holding a horse's bridle. He was so wrapped up in his own thoughts he had not noticed that the horse – which should have been attached to the bridle – had gone its own way. Isaac's mother despaired of her son. He would never make a farmer she decided, so after discussing the matter with her relatives and with the schoolmaster at Grantham she agreed to let him try for entry to Cambridge where he could train to become a country parson.

From Farmhand to Scholar

Isaac Newton was sent back to school to prepare for a university education. He worked hard and easily qualified for entrance to Trinity College at Cambridge. In the autumn of 1661 he arrived to claim his place at the university. At first he had some problems integrating into undergraduate life, but he was fortunate to meet up with another student called John Wickens. The two decided to share a room together, and Wickens was happy to put up with the absent-minded Newton and to assist with the many experiments conducted by his room-mate.

Newton obtained his Bachelor of Arts degree, but his interests ranged far outside the curriculum and he did not pass with distinction. He was able to remain at university, however, to study for a Master's degree. In 1665 a great plague broke out in London and hundreds of people were dying by the week. Cambridge University took no chances; the authorities closed the colleges down for fear of the plague and the students were sent home to fend for themselves. Newton ended up spending most of his time at his mother's rural home. The enforced seclusion in the countryside seemed to have a beneficial effect on him, however, and without the distractions of the university he became totally absorbed in his own ideas and able to continue with his many experiments.

Experiments with Light

In one experiment, conducted in 1666, he acquired a glass prism and passed a ray of sunlight through it. The prism split the sunlight into the colours of the spectrum. These colours had been witnessed by many before him, but Newton went on to make great discoveries about the nature of light. He described his feelings in his own words:

> *I procured me a Triangular glass-Prisme, to try therewith the celebrated Phenomena of Colours. And in order thereto having darkened my chamber, and made a small hole in my window-shuts, to let in a conven-*

ient quantity of the Sun's light, I placed my Prisme
at his entrance, that it might thereby be refracted to
the opposite wall. It was at first a very pleasing diver-
tisment, to view the vivid and intense colours produced
thereby; but after a while applying myself to consider
them more circumspectly, I became surprised to see
them in an oblong form; which according to the received
laws of Refraction, I expected should be circular.

As a result of his experiments with light, Newton discovered the reason why the telescopes of his time always seemed to produce coloured fringes around the image. He realized that the problem could be avoided by making a telescope that used a mirror, rather than a lens, to collect and focus the light. He actually constructed a telescope on these principles, and he was so pleased with his instrument that he sent it to the Royal Society in London for their perusal. This, and his treatise on light, both represented a step forward in the science of astronomy and the Royal Society were sufficiently impressed by these contributions to welcome Isaac Newton as one of their members.

Gravity and Mechanical Matters

But the nature of light was only one of Newton's amazing findings during the years 1665 and 1666. The story that

Newton's theory of gravitation was inspired by the fall of an apple seems to be apocryphal, yet it must be founded in truth, for in old age he told the story to his first biographer William Stukeley (1687–1765) who related it as follows:

> *After dinner, the weather being warm, we went into the garden and drank tea, under the shade of some apple trees, only he and myself. Amidst other discourse, he told me, he was just in the same situation, as when formerly, the notion of gravitation came into his mind. It was occasion'd by the fall of an apple, as he sat in a contemplative mood. Why should that apple always descend perpendicularly to the ground, thought he to himself. Why should it not go sideways or upwards, but constantly to the Earth's centre? Assuredly the reason is, that the Earth draws it. There must be a drawing power in matter: and the sum of the drawing power must be in the Earth's centre, not in any side of the Earth. Therefore does the apple fall perpendicularly, or towards the centre. If matter thus draws matter, it must be in proportion of its quantity. Therefore the apple draws the Earth, as well as the Earth draws the apple. That there is a power, like that we here call gravity, which extends itself thro the universe.*

Newton contemplated the force of gravity. He was convinced that every object in the universe had a gravitational attraction for every other object, and he felt sure that this force was governed by an inverse square law. The apple on the tree was attracted by the gravity of the Earth. Was the Moon in her passage across the sky governed by the same rule? Was the Moon drawn to the Earth by the same gravity as the apple? Newton did some calculations. He was not sure about the size of the Earth, but he found that the calculations fitted the theory *'pretty nearly'*. He was also unsure about which measurement he should take with regard to the distance from the apple to the Earth. Was it the few feet from the ground that he could easily measure? Was it the distance to the centre of the Earth? Was it the distance to an unknown point somewhere under the Earth? He did not know the answer but he kept his calculations and his thoughts for future reference.

He also began to think about mechanics. He thought about the nature of force and acceleration. He wondered about how they were related and whether the laws of mechanics were the same on Earth as in the heavens. Why did the heavens enjoy perpetual motion when everything on Earth ran down because of friction and air resistance? The mathematics to answer his questions did not exist. The equations involved quantities in a state of flux, and he developed a method called 'fluxions' to handle the problems. He had invented what we now call calculus. Newton himself described his years at Woolsthorpe:

In the beginning of the year 1665 I found the Method of approximating series and the Rule for Reducing any dignity of any Binomial into such a series. The same year in May I found the method of Tangents of Gregory and Slusius, and in November had the direct method of fluxions and the next year in January had the Theory of Colours and in May following I had entrance into ye inverse method of fluxions. And the same year I began to think of gravity extending to the orb of the Moon, and having found out how to estimate the force with which a globe revolving within a sphere presses the surface of the sphere, from Kepler's rule of the periodical times of the planets being in a sesquialterate proportion of their distances from the centre of their orbs I deduced that the forces which keep the planets in their orbs must [be] reciprocally as the squares of their distances from the centres about which they revolve: and thereby compared the force requisite to keep the Moon in her orb with the force of gravity at the surface of the Earth and found them to answer pretty nearly. All this was in the two plague years of 1665 and 1666, for in those days I was in the prime

of my age for invention, and minded mathematics and
philosophy more than at any time since.

His mathematical genius was soon appreciated at
Cambridge, and at the recommendation of his tutor Isaac
Barrow (1630–77), Newton was offered the seat of the
Lucasian Professor of Mathematics. For a time his income
was secure. He retired into his ivory tower and worked
on his many other interests, in particular theology and
alchemy. By 1666 he had solved the main mathematical problems relating to gravitation, but he never
prepared his work for publication, and for nearly 20
years these momentous ideas about the universe existed
only in his head.

What is Gravity?
Gravity is the force that draws everything to the ground
– a fact that was well known to people long before Isaac
Newton was even born. However, Newton succeeded in
explaining how the force of gravity works, not just on
Earth but throughout the whole universe. In the 1660s,
in his garden at Woolsthorpe, Newton observed an apple
falling to the ground. He began to wonder if the force on
the apple was the same as the force that held the Moon
in orbit around the Earth. His calculations led him to
believe that it probably was, although it was nearly 20

years before he published his ideas. The force acting on the Moon and an object such as an apple are the same, and Newton was able to assert that every particle of matter in the universe exerts a gravitational attraction on every other particle of matter in the universe.

The great physicist Albert Einstein (1879–1955) developed the theory of mechanics and gravitation beyond the ideas of Newton, but for practical purposes there is no difference between the two concepts.

The Problem of Gravitation

It was December 1683 and in London there was a minor epidemic of the smallpox. It was a bleak and cold Christmas and the smoke from the coal fires left a great smog over the shivering city. The sluggish waters of the River Thames cooled towards freezing point as the sharp frosts of an exceptionally hard winter set in. The narrow arches of Old London Bridge restricted the river water on its passage to the sea, and as the winter approached the ice formed on the slow-moving waters of the Thames above the bridge. It soon became thick enough to support the weight of a man.

On New Year's Day the river had become completely frozen from one bank to the other, and the foolhardy were boasting about crossing the ice to the other side. On 9 January the diarist John Evelyn (1620–1706) crossed over the ice himself. By this time the Londoners felt so safe on the frozen surface of their river that even a coach and horses could make the crossing in safety and whole new streets were appearing on the river. John Evelyn described the scene:

> The frost still continuing more and more severe, the Thames before London was planted with bothes in formal streetes, as in a Citty, or Continual faire, all sorts of Trades and shops furnished, and full of Commodities, even to a Printing presse, where the People and Lady's tooke a fansy to have their names Printed and the day and yeare set downe, when printed on the Thames ... Coaches now plied from Westminster to the Temple, and from several other stairs too and fro, as in the streetes; also on sleds, sliding with skates; There was likewise Bull-baiting, Horse and Coach races, Pupet plays and interludes, Cookes and Tipling, and lewder places; so as it seem'd to be a bacchanalia, Triumph or Carnoval on the Water, whilst it was a severe Judgement upon the land: the trees not onely splitting as if lightning-strock, but Men and Cattel perishing in divers places, and the very seas so locked up with ice, that no vessels could stirr out, or come in: The fowl Fish and birds, and all our exotique Plants and Greenes universaly perishing; many Parks

of deere destroied, and all sorts of fuell so deare that
there were great Contributions to preserve the poore
alive; nor was this severe weather much lesse intense
in other parts of Europe even as far as Spaine, and
the most southern tracts ...

In spite of the severe weather the Royal Society still managed to hold its regular meetings, albeit poorly attended by reason of the extreme cold. The problem of gravitation was a topic very much under discussion, and the astronomers were convinced that the Sun somehow exerted an attractive force on the planets. In Holland Christiaan Huygens (1629–95) had recently published a formula for what he called the centrifugal force – this was the outward thrust experienced by a body such as a planet moving uniformly around a circle. (It was the same outward force that is experienced by a child riding on a roundabout.) This formula for centrifugal force, the astronomers knew, was one of the key factors required to solve the problem of the planetary orbits.

After one of the meetings of the Royal Society three of the members, namely Edmond Halley (1656–1742), Christopher Wren (1632–1723) and Robert Hooke (1635–1703) retired to a coffee house to discuss between themselves the problem of gravitation and how it related to what they called *'the system of the world'*. At this time Wren was turned 50 and Hooke was in his late 40s. They knew each other from their days at Oxford after the Civil War. Halley was the junior member; he was only 26. All three of them had come to the conclusion that the law of gravity was an inverse square law. All three knew that this hypothesis could not be raised to the status of a law unless Kepler's laws of planetary motion, which by now were accepted as established fact, could be proved from it.

But the required proof was a very intricate and taxing mathematical problem. Christopher Wren had tried to solve the problem and he had to admit failure. Edmond Halley was a very able mathematician but he also admitted that his attempt had failed. Perhaps it was a problem that could never be solved. Robert Hooke was not a man to admit failure, however; he coyly announced that he had solved the problem but that he would keep his solution to himself so that others trying to solve it would know the value of his work. Christopher Wren did not believe his friend Hooke had a solution, however, and he offered a book to the value of 40 shillings to anybody who could solve the problem.

A few weeks later it became clear to both Wren and Halley that Hooke's demonstration was not forthcoming. If the problem were to be solved then there was but one person in the whole of England with the mathematical skills to do it, and they all knew that he lived at Trinity

College in Cambridge. Somebody must travel to Cambridge and confront Isaac Newton with the problem.

A Problem Solved

Robert Hooke was definitely the wrong man to make the visit. Newton had taken offence when Hooke had exposed an error in his work to the Royal Society, and Hooke's credibility with Newton was at rock bottom. Christopher Wren was the obvious man to go to Cambridge. First, Newton thought very highly of Wren's skill as a mathematician. Second, while at Cambridge, Wren could witness the progress on the library at Trinity College which he had designed. But Wren was too busy rebuilding London after the Great Fire to spare any time uncovering the secrets of the universe. It was the younger man, Edmond Halley, who had to visit Cambridge and gain an audience with the Lucasian Professor of Mathematics.

Edmond Halley had not met Newton before, but he knew of his awesome reputation. He also knew of Newton's desire for privacy and that he did not suffer fools gladly. Halley was in no hurry to go to Cambridge, and by the time he made the journey in August 1684 the hard winter was history. In the English countryside it was harvest time, and the corn was being reaped with sickle and scythe by every available farmhand in the rural villages. When Halley arrived at Cambridge he had no problem in finding the

Lucasian Professor of Mathematics. He was greeted by a figure with a sharp nose, a prematurely grey but full head of hair and slightly protruding and deeply penetrating eyes. It was Isaac Newton himself. Halley had little notion about how Newton would respond to his request, and he must have been pleased when Newton turned out to be friendly and co-operative. After a little small talk and a few formalities Halley came to the point of his visit by asking the all-important question regarding the system of the world. What did Isaac Newton think would be the orbit of a planet around the Sun, supposing the attraction of the Sun to be *'reciprocal to the square of its distance from it'*? Newton replied immediately that it would be an ellipse. Halley asked him how he knew the answer. *'Why,'* said Newton, *'I have calculated it.'*

Edmond Halley was amazed. He asked Newton for his calculation. The absent-minded professor obligingly began to rummage through his papers as Halley watched with trepidation. Newton was unable to find his paper, but Halley did not doubt for a moment that the man before him was telling the truth. Isaac Newton had already solved the key problem at the heart of the theory of gravitation! Torn between joy and dismay Halley watched in awe as he realized that the man before him carried one of the great secrets of the universe in his head! But try as he might Newton was unable to find the paper with the written solution. It was small wonder that Halley's next act was to ask Newton to rework his calculation.

Publishing a Lifetime's Work

It says much for Halley's tact and diplomacy that he returned from Cambridge with a promise from Newton to renew the necessary calculation and to supply Halley with a copy. It leaves a few questions unanswered, however. Was Newton's study a disorganized chaos of papers and half-completed experiments, or was it as neatly organized as his mind? Perhaps Newton knew exactly where his calculation was but wanted time to think about it before showing it to Halley. Why was Newton, who was always complaining about the demands on his time, so co-operative with Edmond Halley? Newton was well aware of Halley's astronomical work in the southern hemisphere, and at their meeting the two must have discussed many topics in which they had a common interest. Halley told Newton about his findings in St Helena, where his pendulum clock ran more slowly at the top of a mountain. Newton had read the account already in the *Philosophical Transactions,* but there was no substitute for talking to the author face-to-face. Newton probably expounded his ideas on the comet of 1680–81, and he set Halley thinking about the motion of comets and the possibility that they could return after a number of years. The outcome was that when Halley left Cambridge he became progressively more convinced that Newton's ideas must be written up and published for the world to see and read. He knew that the Royal Society was by far the best body to handle the publication.

Halley was very fortunate. He arrived at Cambridge at a time when Newton's researches on alchemy were getting nowhere and his academic life was progressing towards a dead end. The latest discoveries on gravitation had given Newton some new ideas to work on, and he realized that his golden opportunity had arrived and that the time was right to make his researches known to the world. He decided it was his duty to write up and publish these ideas for posterity. He knew they would generate controversy in the world of natural philosophy, but for once he was prepared to publish them and face the consequences.

It took Isaac Newton about a year to put his great work together, which he called *Philosophiae Naturali Principia Mathematica* (*The Mathematical Principles of Natural Philosophy*), usually simply known as *Principia*. Much of it was a synopsis of the problems he had solved over the previous 20 years. He started by formulating his three laws of motion, which today form the basis of mechanics and dynamics. Newton's three laws of motion describe the way bodies move when acted on by a force. The first law states that a body remains at rest or in uniform motion in a straight line unless acted on by a force. We know that moving bodies on Earth slow down and stop, but this is because they are acted on by the forces of friction and wind pressure. If these forces were removed the bodies would continue to move in a straight line forever unless an external force acted upon them.

The second law states that a force changes the motion of a body in the direction of the force. The acceleration of the body depends on its mass and the value of the force. A good example is a train moving in a straight line with a constant force. In such conditions it generates uniform acceleration. For a planet moving around the Sun and drawn to it by gravity, the law is complex but still holds true.

The third law states that for every action there is an equal and opposite reaction. If you sit on a chair, your weight becomes a downward force on the chair. The chair exerts an equal upward force, known as the reaction, on you. When two billiard balls strike each other the result is more complex, but Newton's law shows that the balls experience equal and opposite reactions. These laws appear to apply only to earthbound objects, but Newton went on to show how the same laws could be applied to the motion of the planets around the Sun. He showed how the whole system of the Sun and the planets could be explained from a single law – the law of universal gravitation. Newton used his method of fluxions to arrive at his results, but he reworked them into a classical form so that mathematicians could understand them more readily. The result was a work explaining the whole system of the world in detail, but encompassed in classical mathematics. Future generations would use calculus to solve the problems, leaving Newton's classical proofs in an isolated time warp of their own.

It was 1687 when Newton's great work *Principia* was published and his astronomical works became known to the world. He is remembered as a great scientist and philosopher as well as an astronomer. He retained an interest in astronomy until his death in 1727, and during his lifetime he saw many advances in both the practical and theoretical sides of astronomy.

A Homely View of a Genius

Finally, we must thank Humphrey Newton (fl. 1629–95), a distant relative and assistant to Isaac Newton, for leaving us with this simple but fascinating picture of a master scientist and his often distracted thoughts:

> *When he has sometimes taken a turn or two [about the garden], has made a sudden stand up, turn'd himself about, [he would] run up the stairs like another*

Archimedes, with an 'eureka' fall to write on his desk standing without giving himself the leisure to draw a chair to sit down on. At some seldom times when he designed to dine in the hall, would turn to the left hand and go out into the street, when making a stop when he found his mistake, would hastily turn back, and then sometimes instead of going into the hall, would return to his chamber again.

10

ENGLISH AND
FRENCH RIVALRY

The Royal Society – the national academy of science – was founded in England in 1660 with the purpose of discovering the truth about scientific matters through experiment. In about 1663 some scientists began to hold regular private meetings in Paris, and in 1666 French government minister Jean-Baptiste Colbert (1619–83) was instrumental in formally establishing the group. It gained royal approval, becoming known as the Académie Royale des Sciences. The mission of the Académie Royale des Sciences was to study ways in which the sciences could be exploited to the advantage of the kingdom.

Jean-Baptiste Colbert was also founder member of the Académie des Inscriptions et Belles-Lettres, an establishment set up earlier to choose inscriptions for medals and monuments celebrating the military victories of the Sun King, Louis XIV (1638–1715). (A third foundation, the

Académie Royale d'Architecture, was set up in 1671. Its purpose was to lay down the rules and refine the taste of formal French architecture.) Two French scientists who prepared the ground for the Académie Royale des Sciences, but who died before it was founded, were Marin Mersenne (1588–1648) and René Descartes (1596–1650). In 1611 Mersenne joined the Roman Catholic mendicant Order of Minims in Paris, and from 1614 to 1619 he taught philosophy at the Minim convent at Nevers. He was an ardent opponent of the pseudoscientific doctrines such as alchemy, astrology and related arcane arts, and this was mainly why he vigorously supported true science. He defended the philosophies of René Descartes and the astronomical theories of Galileo (1564–1642). He taught philosophy at the convent L'Annonciade in Paris, and from 1620 onwards he travelled throughout western Europe. Pierre Gassendi (1592–1655) was another French astronomer who did not live to see the foundation of the Académie Royale des Sciences. In about 1614 Gassendi received a doctorate in theology at Avignon and was ordained in the following year. He was persuaded by Mersenne to abandon mathematical and theological pursuits, but he took up astronomy and in 1631 he was the first to observe a transit of the planet Mercury across the face of the Sun.

Sharing Scientific Knowledge

One of Mersenne's most important contributions to science was his long service as a communicator between philosophers and scientists throughout Europe. In his time there were no published scientific journals to distribute knowledge, and sometimes scientists such as astronomers could work for a lifetime on the same project without even knowing the existence of other workers in the field. Mersenne met many scientists on a regular basis and he corresponded at length with Descartes, Girard Desargues (1591–1661), Pierre de Fermat (1601–65), Blaise Pascal (1623–62) and Galileo. Mersenne was such a good communicator it was said that to inform him of a discovery meant to publish it throughout the whole of Europe.

From the 1660s onwards there was always great rivalry between the astronomical and other scientific establishments of England and France. It made for a competitive spirit, and it was responsible for many creative advances in both countries. In most cases the knowledge was published and willingly shared with similar scientific establishments in other developed countries such as Italy, Denmark, Holland and Germany. It was common for members of the French academy to travel to London to meet the Royal Society, and it was equally common for English scientists to make the return trip to Paris. The French had one advantage over the English, however, because for most of Europe Paris was more accessible than London, and so scientists from countries such as Italy, Germany and the Netherlands were more common at their meetings.

The dissemination of information meant that fierce arguments sometimes arose when conflicting theories were published. For example, when Isaac Newton (1642–1727) published his *Principia* in 1687 he wrote a whole section of his book to discredit the views of the French mathematician and philosopher René Descartes, whose work on astronomy was carried out a generation before Newton's. Descartes suggested that the planets were carried around the Sun by a system of vortices, analogous to the swirling motion of water draining into a hole. Newton studied the motion of liquids and he was able to prove, to his own satisfaction, that the motion of the planets could not be explained from the laws of fluid dynamics and that Descartes' system was therefore wrong.

The Great Observatories

A second major contribution to science, and to astronomy in particular, was the French Observatoire de Paris, or Paris Observatory. It was founded under the direction of the Académie Royale des Sciences and became the national astronomical observatory of France. Also like the Académie Royale des Sciences, it was founded by Louis XIV at the instigation of Jean-Baptiste Colbert, and construction began in 1667. The observatory made many important advances in astronomy.

The French, like the English, were very keen to solve the problem of measuring longitude at sea. Gian-Domenico Cassini (1625–1712) studied the satellites of Jupiter and calculated the exact time when the moons were eclipsed by the shadow of the planet. Tables of the eclipses would enable navigators to use Jupiter as a clock in the sky and to tell the time at sea. The method worked well on land, but it was simply not practical to train a long telescope onto Jupiter's moons from the swaying decks of a vessel at sea. The English solution was little better. At Greenwich it was decided to use the position of the Moon to determine longitude. The problem with this method was not the observation itself, but that of predicting the lunar motion as well as the need to make numerous observations before a set of reliable tables could be published.

In the 17th century neither the Greenwich Observatory nor the Paris Observatory solved the longitude problem, but both places made significant contributions to astronomical knowledge. At Paris the astronomer Jean Picard (1620–82) measured the length of a degree of latitude with greater accuracy than any before him, helping to establish the size of the Earth. In 1672 Picard and

Cassini made the first realistic measurement of the astronomical unit, the distance from the Earth to the Sun. This was achieved by measuring a parallax for the planet Mars when it was at its closest approach to the Earth. John Flamsteed (1646–1719) and Richard Towneley (1629–1707) made the same observation in England, but the French were able to publish their results first and claim priority.

Measurements of Light

In 1676 another very important astronomical constant was measured for the first time. It was the speed of light. The measurement of the velocity of light was one of the great triumphs of the Paris Observatory and it came about almost as an accident. The Danish astronomer Ole Rømer (1644–1710) was working on the longitude problem and was trying to create a set of tables for viewing the satellites of Jupiter. He was puzzled to discover that when Jupiter moved further away from the Earth the times of the eclipses were measurably later than his tables predicted. After some consideration he correctly concluded that the delay was due to the finite velocity of light. It took about 40 minutes for light to travel from Jupiter to Earth, but this time varied according to the distance between the two planets, and the variable distance had not been allowed for in the calculations. Thus, when the Earth and Jupiter were on the same side of the Sun, the distance between them was far less than when they were on opposite sides of the Sun. When he realized that the light therefore sometimes had further to travel, with the extra distance being as much as the diameter of the Earth's orbit, Rømer was able to calculate the speed at which light travelled from Jupiter to the Earth, and hence the speed of light itself. This was long before Albert Einstein's (1879–1955) discoveries about the speed of light; in Rømer's time nobody knew that the speed of light was the maximum achievable speed, but it was well known how important this constant was to physics and astronomy.

At the beginning of the 18th century there seemed to be very little progress in astronomy. In 1704 Newton at last published his treatise on light, but he had waited for the death of his adversary Robert Hooke (1635–1703) before going to print. Was light a wave motion or did it consist of small particles? Newton was unable to come up with a direct answer to this question. Some experiments suggested a wave motion, but others suggested that light consisted of particles. Newton knew about interference fringes; these were easily explained by the wave theory of light, but not by the particle theory. He was also puzzled as to how light could travel across the distances between the stars. This was easy to explain with the particle theory, but he wondered about how a wave could travel through the vacuum of empty space where there was no air or other medium to carry it.

The Newtonian model of the universe stood up well to most of the questions about the motions of the planets and the stars, but there was much speculation about the nature of the universe. How far away were the stars? It had been difficult enough to measure parallax for the planets, but there seemed no way to measure the distance to the stars except by the indirect method of assuming that they were bodies with the same brightness as our own Sun. At this time a few astronomers believed that the Sun was a star, but most still believed the Sun to be the largest object in the universe with the stars as much lesser bodies. They could not believe that the universe was so large that a star as bright as the Sun could appear as a tiny dot of light in the sky.

Some Great Conundrums

There were other questions being asked about the universe. Why were the stars not drawn together by gravitation? Some suggested that perhaps they were drawn together, and at some time in the future the universe would end in a massive implosion. The theologians calculated from the genealogies in the Bible that the world was nearly 6,000 years old, and from the Book of Revelation they calculated that it would end in about the year 2000. Was the universe infinite? Later, an astronomer called Heinrich Olber (1758–1840) posed that, if the universe was indeed infinite, but was filled with a constant density of stars, then the night sky should be as bright as the surface of the Sun. The sky would be bright everywhere, not black with tiny spots of light. The problem became known as Olber's paradox. Some suggested that perhaps the stars were contained in a great disc, all of them orbiting a common centre, rather like a scaled-up version of the solar system. This was nearer to the reality, but still a long way away from the truth.

There was another dilemma put forward in Newton's lifetime. It involved the force of gravity. Basically every star or planet has what is called an escape velocity. Any orbiting object travelling at this speed can escape from the gravitational pull of the star or planet. For example, an object travelling at above 6.8 miles per second (11 km/sec) with no air resistance can escape from the Earth. It needed a higher velocity to escape from the Sun, but comets were discovered with velocities high enough to make their escape and they were never seen again. The escape velocity from a star of any given mass could easily be calculated by classical mechanics. The dilemma was that if a star was sufficiently massive then the escape velocity could be greater than the velocity of light. It meant that even light itself could not get away from such a massive star. It was an idea first surmised by John Michell (1724–93) in the 18th century, but for over 200 years it was no more than a fanciful concept, for such a vast object had not been, nor could be, observed. What Michell had envisaged, however, was the phenomenon later to be known as a black hole.

Calculus 'Wars'

In Britain, mathematics had been given a great boost by the publication of Newton's *Principia*. Although he had developed the calculus to arrive at many of his conclusions, he chose not to use the methods in the *Principia* for he feared that nobody would be able to understand them. It happened that at the same time as Newton was developing calculus there was another mathematician, Gottfried Leibniz (1646–1716) of Germany, who was working along the same lines. In spite of warnings from people like John Wallis at the Royal Society, Newton persistently refused to publish his work on calculus and the result was that Leibniz published the discovery of calculus before him. There followed a great feud as both parties tried to establish their priority, which did little to enhance the image of either Newton or Leibniz. One result of the dispute was that the British used a different notation from the rest of Europe. In the 18th century, after the death of Newton, the continental notation was the one that came into general use and British mathematicians found themselves out of step with the rest of Europe. Thereafter, British mathematics fell into a decline, and towards the end of the century the key developments came from the Continent and from the French in particular.

The leading mathematicians of the century all wanted to make a contribution towards the theoretical aspects of astronomy. The Swiss mathematician Leonhard Euler (1707–83) was born at Basel. He worked on analytical geometry and the theory of complex numbers, but possibly his greatest work was on the mechanics of rotating bodies. He showed how to calculate their motion from their three principal axes and moments of inertia. He did much work on calculus and he tried to solve the three-body problem that had defeated Newton in the previous century. Euler was the most prolific mathematician of the century, but he was closely challenged in this field by two French mathematicians, Joseph-Louis Lagrange (1736–1813) and Pierre Simon Laplace (1749–1827).

Joseph-Louis Lagrange was born in Italy to a French family. He moved to Paris and in 1764 won an academy prize for his essay on the libration of the Moon (the small oscillation of the Moon about its mean position). His greatest work was his *Méchanique analytique* published 1788, in which he showed how to formulate mechanical problems in terms of generalized coordinates of position and momentum.

Pierre Simon Laplace was the son of a peasant farmer in Normandy. Laplace was a precocious child. He quickly left his agricultural roots behind him and arrived in Paris at the age of 17 to study mathematics. He worked on gravitational problems and the anomalies of the planetary orbits. He was able to show that the errors in the orbits of Jupiter and Saturn could be accounted for by the gravitational attraction between them. His great work was the *Méchanique celeste;* it consisted of five volumes published between 1798 and 1827.

Throughout the 18th century it was the French who produced the best mathematical theories of the mechanics of the universe. Lagrange's equations were developed to solve mechanical systems, and Laplace developed a very powerful tool for solving problems related to gravitational fields. Although the British lagged behind in mathematics at this time – relying instead on Newton's *Principia* – in terms of solving practical problems such as finding longitude and building larger and more powerful telescopes, they would continue to make great strides forward.

A Family of Astronomers

Italian-born French astronomer Gian-Domenico Cassini (1625–1712) was the first of four generations of his family to hold the post of director of the Paris Observatory. He discovered the true nature of the rings of Saturn, and his telescope was powerful enough to see the gap between the rings that became known as the Cassini division. He also discovered four of Saturn's satellites and made an accurate estimate of the distance to the Sun. Amongst his other achievements, he was first to record observations of the zodiacal light and he laid down three rules that accurately describe the rotation of the Moon.

Jacques Cassini (1677–1756) succeeded his father Gian-Domenico as head of the Paris Observatory in 1712. He compiled the first tables of the orbital motions of Saturn's satellites, and in 1718 he completed the measurement of the arc of the meridian – the line of longitude passing between Dunkirk and Perpignan. In his paper *De la grandeur et de la figure de la terre*, published in 1720, he argued that the Earth was not a true sphere but was elongated at the poles.

Cassini de Thury (1714–84), sometimes known as Cassini III, continued the surveying work undertaken by his father Jacques, and he began the construction of a great topographical map of France. He, his father and his grandfather had defended the Cartesian view that the Earth was slightly elongated, but Cassini de Thury abandoned the position in the face of growing evidence that favoured the opposite – the so-called Newtonian view that the Earth is flattened at the poles. Cassini de Thury succeeded his father as director of the Paris Observatory in 1771. The *Carte géométrique de la France* (*Geometric Map of France*), or *La Carte de Cassini*, was the first map of an entire country drawn up on the basis of extensive triangulation and topographic surveys. It was published in 1789, the year of the French Revolution.

Jean-Dominique, comte de Cassini (1748–1845) succeeded his father as director of the observatory in 1784. He completed his father's map of France, which formed the basis for the *Atlas national* of 1791, depicting France's departments.

11

FINDING LONGITUDE

Mapping the world using a grid of intersecting circles was first implemented in the ancient world. The astronomer Ptolemy was not the first to suggest that the Earth was a sphere, but we must give him the credit for suggesting that the grid of lines we now know as latitude and longitude was the best way of mapping a set of points on such a surface. Thus one of the greatest requirements facing geographers and explorers was the need to determine the latitudes and longitudes of points on the surface of the Earth.

Even in the ancient world, distances north and south of the equator could easily be calculated by observing the Sun. The measurement of the elevation of the midday Sun above the horizon was all that the navigator needed to find his latitude. The calculation was affected by the various seasons, but it was a relatively straightforward task to build in a correction factor to allow for the time of the year. At night it was even easier to find latitude; it was simply a matter of measuring the height of the Pole Star above the horizon. With the invention of the astrolabe, the identification of any known star enabled the latitude to be calculated. The only time that latitude could not be found was when prolonged cloud cover meant that none of the heavenly bodies was visible.

In Sight of the Land

Distances around the Mediterranean could be measured directly along the coastlines with comparative ease simply by pacing out the distance, or sometimes by triangulation. Finding the longitude was therefore not a serious problem so long as land was in sight. It was when the ships began to explore beyond the sight of land or outside the Mediterranean that finding the longitude became a requirement. The problem was not confined to sea voyages; the longitudes of distant places reached by land journeys, such as China and Japan, were sometimes estimated with errors as much as 90 degrees away from the true values. It was small wonder that when Christopher Columbus (1451–1506) sailed to the west he thought the land he discovered lay off the coast of the Indies.

Using the Sun, the Moon and the Stars

Thus finding the longitude, the east–west distance from

a fixed meridian, was a far more difficult task than finding latitude, but it was still essentially a case of knowing the local time and comparing it with the time at a zero meridian. Local time could be calculated from the position of the Sun at noon, but finding the time at the zero meridian was a very different matter. The English calculated longitudes from the Greenwich meridian and the French worked from the Paris meridian. By the 17th century both countries were fully aware that there were astronomical events in the skies that could be used for determining the time at the zero meridian by an observer at another point on the Earth. The most obvious was an eclipse. Astronomers could predict the path of a total solar eclipse and the time at any point along the path. Such an eclipse was a rare event, but an eclipse of the Moon was much more common and was almost as useful. In many ways it was actually better, and it could be seen wherever the Moon was visible. The eclipse was an excellent method for determining longitude on land, but the navigator would need an eclipse every night to help with the problem of longitude at sea.

The English opted for using the position of the Moon to help navigators calculate the time at Greenwich; it was the most prominent object in the night sky and it could easily be observed from a ship at sea. As we have seen, the French developed a different method; they studied the motion of the moons of Jupiter. They reasoned that the moons could be used as a clock in the sky, and if they could produce tables of the regular eclipses of the moons by observing them passing in or out of the shadow of the giant planet, then navigators could use the published tables to calculate the time at the Paris meridian.

The French method of using the Jovian satellites was the simpler of the two approaches. Galileo (1564–1642) was the first to observe the satellites of Jupiter using his recently invented telescope, and he was quick to realize that the satellites kept regular time as they orbited their planet. Tables were constructed to predict the times when the satellites entered the shadow of Jupiter, and Galileo went on to design a special helmet for finding the longitude – it had a telescope attached to one of the eyeholes. The method was not easy to use on land, but on the swaying deck of a ship at sea it was almost impossible to see Jupiter let alone the satellites. The observation was so difficult that even Galileo had to admit that the pounding of the observer's heart could cause the whole of the planet to jump out of the telescope's field of view.

The English fared a little better with their approach. The Moon was the most prominent object in the night sky and it moved across 12 degrees of sky in the course of a day. All that was necessary was to produce tables showing the position of the Moon at any time of day or night. By observing the background stars nearest to the Moon its position in the sky could be calculated. The tables in the nautical almanac could then be used to calculate the time at Greenwich, and hence the longitude. In daytime, if the Sun and the Moon were both visible in the sky, then all that was required was to measure the angular distance between them.

The Greenwich Observatory

In 1675 a decision was made by Charles II and his advisers to build an astronomical observatory in the Royal Park at Greenwich, a village about 5 miles (8 km) from the centre of London. This was an obvious location for the new observatory, for it was sufficiently far from London not to be troubled by the smoke of the city, and it was right at the heart of the shipping being on the banks of the River Thames. There was a small hill in the location where in the distant past there had once stood a Norman castle, and Christopher Wren (1632–1723) thought this was the ideal site. On 22 June 1675 a royal warrant addressed to the Master General of the Ordnance outlined the plans and purpose for the new observatory:

> *Whereas, in order to the finding out of the longitude*
> *of places for perfecting navigation and astronomy, we*
> *have resolved to build a small observatory within our*

park at Greenwich, upon the highest ground, at or near the place where the castle stood, with lodging rooms for our astronomical observator and assistant, Our Will and Pleasure is that according to such plot and design as shall be given you by our trusty and well-beloved Sir Christopher Wren, Knight, our surveyor-general of the place and scite of the said observatory, you cause the same to be built and finished with all convenient speed, by such artificers and workmen as you shall appoint thereto, and that you give order unto our Treasurer of the Ordnance for the paying of such materials and workmen as shall be used and employed therein, out of such monies as shall come to your hands for old and decayed powder, which hath or shall be sold by our order on the 1st of January last, provided that the whole sum to be expended or paid, shall not exceed five hundred pounds; and our pleasure is, that all our officers and servants belonging to our said park be assisting to those that you shall appoint for the doing thereof, and for so doing, this shall be to you, and to all others whom it may concern, a sufficient warrant.

Thus finding a solution to the longitude problem was the main reason for the founding of the Royal Greenwich Observatory. In 1675 John Flamsteed (1646–1719) was appointed as the first Astronomer Royal. The observatory was supplied with several clocks. There were two clocks constructed with a very short swing of the pendulum, one of which was designed to show sidereal time – the rotation of the stars rather than the Sun – which was an essential requirement for such an observatory. There were seven clocks altogether in the observatory, plus a micrometer, a sextant and a mural arc. There were four telescopes with focal lengths that varied from 2.5 to 5 metres (8–16 ft). Flamsteed was given a very meagre allowance for instruments, but he was prepared to put his own money into the observatory. He took a job as the incumbent of Burstow near Reigate, and this provided him with a second income.

Plotting the Motion of the Moon

Flamsteed began to work on a set of accurate observations of the Moon to determine the precise orbit. He was an excellent and meticulous observer, but he soon discovered a problem with the lunar method: it lay not in the difficulty of the observation itself but in the complex motion of the Moon. The orbit of the Moon around the Earth was analogous to the orbits of the planets around the Sun – in other words, it was an ellipse with the Earth at one focus. But the Moon was also influenced by the gravity of the Sun, and so the ellipse could only be an approximation, and its properties varied as the Moon progressed in its orbit. In the 1660s there were a number of people who claimed to have solved the problem. Flamsteed was not convinced by any of them, but he had discovered a text on the lunar motion written by the brilliant young English astronomer Jeremiah Horrocks (1618–41), better known for his observation of the transit of Venus. Flamsteed found Horrocks' theory of the lunar motion to be the best available at the time. It was not a new theory, however, having been developed by Horrocks in the late 1630s just before his death. The great mathematician Isaac Newton (1642–1727) tried to devise a more accurate theory for the motion of the Moon, based on his inverse square law of gravitation. He failed, however, because the problem actually involved the motion of three bodies – the Sun, the Earth and the Moon – and it required a unique solution.

Flamsteed spent the rest of his life working on the problem of the lunar motion, in addition to his other astronomical duties and his job as a parish priest. Flamsteed refused to be hurried and it took him many years to publish his work. He had been measuring the position of the Moon since his investiture as Astronomer Royal in 1676, but by the end of the century he had still not published his findings. There was a controversy over the long delay. Flamsteed wanted to withhold his results until they were complete, but he was taking such a long time that other researchers were delayed and his results were urgently needed by Newton, Halley (1656–1742) and others. Isaac Newton, in his capacity as president of the Royal Society, was able to press for immediate publication. In 1704 Prince George of Denmark undertook the cost of publication. Edmond Halley edited the incomplete observations and 400 copies were printed in 1712. Flamsteed was very angry at the way in which his work had been treated. He managed to purchase 300 of the copies and to ceremoniously burn them. His own version of the star catalogue, the *Historiae Coelestis Britannica*, was published 13 years later in 1725. It listed more than 3,000 stars and gave their positions much more accurately than in any other previous work.

A New Astronomer Royal

In 1720 Edmond Halley succeeded John Flamsteed as Astronomer Royal. Halley is best known for the comet named after him, although he made several other important contributions to astronomy.

Halley spent several years on the Atlantic island of St Helena where he studied the skies, helping to improve the charts of the stars in the Southern Hemisphere. He took a pendulum clock with him and made two interesting discoveries. One was that his clock seemed to run

more slowly at St Helena than it did in the northern latitudes – a fact that he rightly concluded was due to a
lower value of 'g' (the acceleration due to gravity) near
the equator than at the poles. He also took his clock to
the top of a mountain and found to his satisfaction that
it ran a little slower at high altitude; this effect was also
caused by a small change in the value of 'g'. These changes
were caused by centripetal forces and the fact that the
Earth's radius is larger at the equator.

It has to be said that Edmond Halley and John Flamsteed were not the greatest of friends. The main reason
was their differences over religion. Flamsteed was a
devout churchman whereas Halley was an atheist and
made no secret of his views. His atheism very nearly
prevented him from advancing his career. In 1691 he
failed to obtain a professorship at Oxford because of his
unorthodox views. The mathematician William Whiston
(1667–1752) described the event:

> I will add another Thing which I also had from Dr.
> Bentley himself. Mr. Halley was then thought of for
> successor, to be in a Mathematick Professorship at
> Oxford; and Bishop Stillingfleet was desired to recom-
> mend him at Court; but hearing he was a sceptick,
> and Banterer of Religion, he scrupled to be concerned;
> till his Chaplain Mr Bentley should talk with him

about it; which he did. But Mr Halley was so sincere
in his infidelity, that he would not so much as pretend
to believe the Christian Religion, tho' he thereby was
likely to lose a Professorship; which he did accord-
ingly; and it was then given to Dr Gregory: Yet was
Mr Halley afterwards chosen into the like Professor-
ship [of Geometry] there, without any Pretence to the
belief of Christianity.

When Halley returned from his voyage to the Atlantic he could curse as fluently as any seaman. This was also too much for the orthodox Flamsteed.

A Prize for Finding the Longitude

The problem of finding the longitude at sea was proving to be far more difficult than anyone had imagined, and progress was very slow. The Board of Longitude decided to offer a prize of £20,000 – a fortune in the 18th century – to anyone who could build a clock to keep good time at sea or who could solve the problem of finding the longitude by any suitable method. Thus, at the same time as the astronomers at Greenwich were working on the lunar motion, one man was devoting his life to solving the longitude problem by another method. In 1693, at Foulby in Yorkshire, a child called John Harrison was born. His father was a carpenter and the family moved to live at Barrow on Humber in Lincolnshire. As a child, young John was brought up to work with wood, and he was so precise with his woodwork that before the age of 20 he had built a working clock made almost entirely from wooden pieces.

When John Harrison (1693–1776) heard about the Board of Longitude prize he devoted his whole life to winning it. He travelled to London where he managed to get an audience with George Graham (1673–1751), the premier scientific instrument maker of the times. At first Graham did not take Harrison very seriously, but he soon realized that the younger man had many ideas in his head for the new timepiece, and that he knew how to solve the problems involved. Graham ended up being so impressed that he gave Harrison a generous loan, with instructions for Harrison to repay the money when his finances allowed.

The first problem to be solved was to find a new regulating device for a timekeeper. The pendulum was simple and accurate but it could not cope with the swaying of a ship at sea. Harrison devised a movement with oscillating brass weights and springs that kept good time on the moving deck of a ship. He knew that higher temperatures caused a clock to run more slowly because of the expansion of the metal parts. So he experimented using a balance wheel made from a bimetallic strip of brass and steel. It was designed to retain the same radius when the temperature changed.

Harrison built five timepieces in all. His first piece (H-1) was tested in May 1741 on a six-week voyage to Lisbon and back. The outward voyage was made on the *Centurion*, which later became the flagship for George Anson's circumnavigation of the world. The records show that on the return journey Harrison correctly located the ship off the Lizard, whereas the official navigator, Roger Willis, believed the ship to be at Start Point. However, Willis' reckoning was clearly wrong, since Start Point is many, many miles further east, off the Devon coast near Salcombe. However, Harrison was not offered the prize for his timepiece. This was mainly due to jealousy on the part of the Astronomer Royal, Nevil Maskelyne (1765–1811), who wanted to win the prize with his own lunar method. The Board of Longitude was sufficiently interested to provide John Harrison with subsidies, however, and he persevered with his ideas for more than 20 years. He produced a second chronometer (H-2) in 1741, but for some reason this device was never tested at sea.

It took Harrison until 1759 to produce his third chronometer, H-3. The reason why it took him so long was that he kept encountering new ideas and improvements as he was working on it. Harrison's fourth chronometer (H-4) was completed a year later in 1760.

The Board of Longitude decided to test the two chronometers together.

A Stern Test

Eventually, the chronometer H-4 was taken on board HMS *Deptford* and the third chronometer H-3 was taken out of the running. Thus everything depended on H-4. The *Deptford* set sail for Jamaica under Captain Dudley Digges with William Harrison (1728–1815), son of John Harrison, as curator of the chronometer. Out in the Atlantic Ocean a major crisis occurred when it was discovered that the ship's supply of beer was unfit for human consumption. To the dismay of the sailors the beer had to be thrown overboard and they were reduced to drinking water. William Harrison declared that according to the chronometer they would sight the island of Madeira the following day, where they could replenish their stocks of beer. Captain Digges disagreed, however. According to his traditional method of dead reckoning Madeira was still several days sailing away. He was willing to lay a bet on his opinion. The next morning Madeira was sighted. Dudley Digges was a good loser. He was very impressed and offered to buy the next chronometer made by the Harrisons. He wrote to John Harrison with the news:

Dear Sir,

I have just time to acquaint you with the great perfection of your watch in making the island on the Meridian; According to our log we were one degree 27 minutes to the Eastward, this I made by a French map which lays down the longitude in Teneriffe, therefore I think your watch must be right.

Adieu.

Harrison's fourth chronometer was accurate to within an error of only 1 mile (1.6 km) in the test voyage to Jamaica – easily close enough to win the prize offered by the Board of Longitude. On a second voyage to Barbados three years later the watch did not perform quite so well but the error was still less than 10 miles (16 km). In 1765 Harrison was paid only half the £20,000 reward although he had clearly met all the requirements. He only received the other half after a protracted legal wrangle that was settled by a private Act in 1773. Even then it needed the intervention of the king before Harrison was paid. The reason for this was that the Board of Longitude was comprised mainly of jealous astronomers and mathematicians who thought that their own lunar method was the solution to the problem of finding the longitude. And indeed, the astronomical method of finding longitude was at last coming to fruition. By a strange coincidence, the two methods became available at almost the same time, although evidence shows that John Harrison was the winner by a narrow margin. Whilst we must acknowledge the time-consuming effort put in over many years by those seeking to provide an accurate astronomical method for finding latitude, we must also applaud the perseverance and skill of John Harrison.

In the 1750s the major advance in finding the longitude by the lunar method was made when the German mathematician Tobias Mayer (1723–62), guided by the Swiss mathematician Leonhard Euler (1707–83), produced a more accurate theory for the lunar motion. Eventually, in the 1760s, the first nautical almanac was published with instructions for finding longitude from the position of the Moon in the sky – nearly a century after the foundation of the Royal Greenwich Observatory.

An exact duplicate of Harrison's fourth chronometer was made by Larcum Kendall for the Admiralty at a cost of £450. The cost of manufacturing the chronometers reduced quickly and few years later Thomas Earnshaw (1749–1829), the inventor of a modified design, was producing chronometers for less than one-tenth of Kendall's price.

Using the Nautical Almanac
In 1768 when Captain James Cook (1728–79) embarked

on his first voyage of discovery to the Southern Hemisphere, he carried with him a professional astronomer called Charles Green (1735–71). He also carried with him a copy of the nautical almanac published by the Royal Greenwich Observatory for the determination of longitude at sea. Green was one of the few men who could find the longitude from the position of the Moon by using the tables in the almanac. Thus Cook's *Endeavour* became the first ship to enter and cross the Pacific Ocean and to know her longitude on the surface of the Earth. The voyage was also intended to make another important contribution to astronomical knowledge. This was the determination of the Earth–Sun distance – the astronomical unit – from the observation of the planet Venus on the face of the Sun. The transit of Venus was successfully observed from the island of Tahiti, but the observers could not agree about the precise timing of the entry of Venus onto the face of the Sun. The problem was a phenomenon known as the 'tear drop' effect, an optical illusion that showed the shadow of Venus still joined by a 'thread' to the rim of the Sun when in fact it was already into the transit.

12

WILLIAM HERSCHEL
Gazing Deeper into Space

There is a well-known correlation between music and mathematics, amply demonstrated in the account of the life of William Herschel (1738–1822). The son of a German musician, William Herschel was himself a professional organist before becoming fascinated by the stars – a fascination that led him to become one of the greatest of all astronomers and a pioneer in the improvement of telescopes used for watching the night sky.

Isaak Herschel was a musician in a German military regiment. In 1755, when the regiment was transferred to England for several months, Isaak took his family with him. His young son, Wilhelm, also played in the regimental orchestra, but he had not signed up as a soldier and when his father's regiment was transferred he decided to stay in England in the hope of making a living from his musical talents. Wilhelm was successful in this endeavour, and he was employed to play in the Duke of Richmond's private orchestra. He moved to Leeds where he remained for four years, before moving to Halifax for a short time. In December 1766 Herschel was appointed as the organist at the Octagon Chapel in the fashionable city of Bath. Wilhelm Herschel was then 28 years of age, and about this time he changed his Christian name from Wilhelm to the anglicized version of William.

Improving the Telescope

However, it was not as a musician that William Herschel became famous. But it was through the patterns in his music that Herschel became interested in mathematics, and this in turn led him on to an interest in astronomy. Soon he became fascinated by what he could see in the night sky, and he found that he was spending all his spare time studying the stars. Simple telescopes were easy to come by, but after a short time they were not good enough for Herschel's observations. No matter how far into the sky he could see he always wanted something better so he could see further. The local opticians and spectacle makers supplied him with better lenses and eyepieces, but even they were found to be of limited use.

He knew that reflecting telescopes could be purchased from London instrument makers, but they were very expensive and Herschel could not afford one. He also knew that in the previous century Isaac Newton (1642–1727) had made his own reflecting telescope. Herschel knew that the reflecting telescope had many advantages over the refracting telescopes of the time. The mirrors were free from the coloured fringes that appeared in the refracting telescope, a defect known as chromatic aberration. The mirrors were easier to manufacture than lenses, and because they could be supported from the back they could also be made much larger. Herschel's only solution was to set about casting and grinding his own mirrors. It was painstaking work. He cast larger and larger mirrors for his telescopes, but he also met with many failures. The larger the mirror the more likely it was to crack during the cooling process. Even when Herschel had cast a perfect mirror his problems were not over. He had to spend hours and hours polishing the mirrors to achieve the perfect parabolic surface. He mastered this and other techniques, and he also made his own eyepieces. He persevered with larger and larger telescopes year after year, and by the time he had mastered all the optical techniques he was building the best telescopes in the world.

Herschel was ably supported in his endeavours by his sister, Caroline (1750–1848), who was also a musician and a professional singer. Caroline undertook all the household duties at their Bath residence in New King Street, and she made sure that her fanatical brother was properly fed. She also spent hours and hours helping to grind and polish the mirrors to perfection. Without the support of his sister, Herschel would never have become the most famous astronomer of his time.

A New Focus on the Moon and Stars

When his first reflecting telescope was ready for observing, one of the first tasks tackled by William Herschel was to study the Moon. He knew the distance to the Moon, so he could estimate the height of the mountains from the shadows they cast on the surface. He made a survey and he drew maps of his findings. Then he looked deeper into space. His reflecting telescope could see far more stars than were visible with the naked eye and more than were listed in the most up-to-date star catalogues. He became interested in the different types of stars, and he studied objects such as nebulae (gas clouds) that were not stars at all. He soon realized that a large number of stars were not single stars like our Sun but were paired, or binary, stars. Others were triple stars or even clusters of several stars all rotating about each other. He went on to make a detailed study of the binary stars, and he was able to distinguish genuine binary stars from others that looked to be paired or clustered but in reality just happened to lie in the same direction and were in fact light years apart.

With his reflecting telescope William Herschel was able to see much further into the sky than others before him. His catalogue of binary stars eventually grew to 700 entries, and soon the number of known double stars in the sky grew to nearly 10 per cent of the number of single stars like the Sun. He discovered many new nebulae to add to the French astronomer Charles Messier's collection, and the extended catalogue added up to about 250 objects. His sister Caroline was keen to do more than look after the house and polish telescope mirrors and so he supplied her with one of his telescopes. It was not long before Caroline Herschel had also discovered new objects to add to the Messier catalogue, plus binary stars and several new comets.

The Herschels became well known in Bath, and sometimes received visits from local residents and others who wanted to talk about astronomy and look through the telescope. One of the visitors was none other than Nevil Maskelyne (1732–1811), the Astronomer Royal and director of the Royal Greenwich Observatory.

An Amazing Discovery

The Herschels made excellent progress as they continued to map the skies and build up catalogues of new objects. Both were fanatical about their work and for years they were happy to be making additions to the catalogues. Then, on 13 March 1781, William Herschel made a discovery that was every astronomer's dream. For some time he had been watching a star that he suspected was moving against the background of the other stars. It had no tail and so it was certainly not a comet. Nor was it listed in the Messier catalogue. After more observation Herschel became convinced that the object really was moving, and further- more he was also convinced that it was moving in a nearly circular orbit. There seemed to be only one possible expla- nation for the motion. William Herschel had discovered a new planet!

The discovery of a new planet launched Herschel's career. The invention of the telescope had paved the way for countless additional stars to be detected by astronomers. Many new comets and even other galaxies had been discovered, but nobody had found a new planet since earliest times. It was thought that all the planets had been discovered at the dawn of astronomy, and thousands of years of gazing at the heavens since then had not changed this belief. When the position and direction of motion of Herschel's new planet had been measured it was easy for other astronomers to verify the find. As the path of the new object in the sky was plotted, there was no doubt that it was further from the Sun than Saturn. After a time it became obvious that it was obeying Kepler's and Newton's laws, and it was following an elliptical course around the Sun. Now all that remained was to name the new planet. Some suggested that it be called planet Herschel. Others suggested the rather pompous name of Georgium Sidus to honour King George III, but after his incompetent handling of the American Revolution George was not the most popular of kings.

In the best classical style the name Uranus was eventually agreed upon. A few years later Herschel discovered two moons orbiting Uranus and they became known as Titania and Oberon, thereby seemingly disclosing his taste for the plays of William Shakespeare. William Herschel seems to have been too modest to name the moons himself and the names were actually given many years later by his son John.

Some would say that the discovery of Uranus was the high point of Herschel's career, and in the sense that it brought him fame and fortune this is true. He was given a pension by the king and he moved to Berkshire where he was able to set up his own observatory. It was only then, when he had royal sponsorship, that he was able to make his most significant contribution to the advance of astronomy. In 1785 he and his sister Caroline moved to a house called Clay Hall in Old Windsor. Socially it was a very acceptable address, but the Herschels did not own the property and in 1786 they were obliged to move to a new residence in nearby Slough. His dwelling became known as Observatory House. He found Slough to be very amenable, and it was there that he met his wife, a widow called Mary Pitt. On 7 May 1788 they were married at St Laurence Church in Slough. William's sister Caroline moved into nearby lodgings and continued her work as his assistant.

Herschel's Biggest Telescope

During his long career it has been estimated that William Herschel made about 400 telescopes. His salary as the court astronomer was supplemented by an income from the sale of his telescopes. He became so well known as a telescope maker that his customers included the king of Spain, the prince of Canino and the Russian emperor. He made telescopes for export to Berlin and he even sold one of his instruments to China.

The largest and most famous was his great reflecting telescope with an aperture of 1.2 metres (4 ft) and a focal length of 12 metres (40 ft). The telescope was completed in February 1787 but Herschel was not satisfied with the mirror; it was too thin and it bent under its own weight of half a ton. He ordered a new disc to be cast but it cracked in the cooling process. Then he cast a third mirror 8 centimetres (3 in) thick and weighing nearly a ton. It was polished and ready for use in August 1789. It was with this telescope that Herschel discovered two moons of Saturn within the first month of observation. They became known as Mimas and Enceladus and, as with the moons of Uranus, it was his son who chose the names long after William's death. The discovery of the moons took place just one month after news arrived in England of the storming of the Bastille in Paris and the outbreak of the French Revolution.

The great telescope needed a complex system of scaffolding and pulleys to raise it to the desired elevation. The whole structure was mounted on a large turntable that needed to be winched round to the correct azimuth to point it towards the required object in the sky. This type of mounting made it very difficult to follow a star as the Earth rotated, and consequently the great telescope was only used to follow up discoveries from smaller instruments. The smaller but more manoeuvrable 6-metre (20-ft) telescope was the one Herschel used for most of his observing.

Some Strange Views

Herschel wanted to know more about the distribution of the stars in the sky. To accomplish this he divided the sky into 683 regions and then set about trying to count the number of stars in each region. He produced a model showing the distribution of the stars in the Milky Way. It consisted of a great spinning disc in which he placed the Sun, our own star, near the centre. His model of the galaxy was the most advanced produced at that time but in fact it contained one of the few mistakes made by Herschel. The Sun is actually situated on one of the spiral arms of the galaxy, quite a long way from the centre where Herschel put it.

Herschel was well acquainted with Newton's work on the spectrum and he was one of the first to realize that there were other colours (or wavelengths) that lay outside the visible spectrum. He passed a beam of sunlight through a glass prism and he held a thermometer just beyond the red end of the spectrum. The thermometer indicated a temperature rise. Herschel came to the conclusion that there must be an invisible form of light beyond the red wavelength. It became known as infrared radiation. Herschel was an excellent scientist but he still held some unconventional views that may have appealed to science fiction readers. He thought, for example, that every planet was inhabited by intelligent life, and he also believed that there was a temperate region beneath the surface of the Sun where there lived a race of beings with very large heads who had fully adapted to their environment.

In Herschel's time nobody had any idea of the true size of the universe. It seemed impossible that anybody would ever be able to measure a parallax for the stars. Some of the 'fixed' stars were not quite in the positions given by Ptolemy and Hipparchus, however, and this gave reliable evidence that they had moved over a time span of nearly 2000 years. The nebulous smudges of light revealed by the large telescope for the objects in the Messier catalogue were thought to be clouds of luminous gas; some of the Messier objects were in fact distant galaxies much further way than the stars, but in Herschel's time no one imagined that they were any more distant than other objects in the sky. In Herschel's time the calculations of the French mathematicians Lagrange and Laplace implied that the Newtonian model of the universe reigned supreme.

Carrying on the Work

In 1816 William Herschel was made a knight of the Royal Guelphic Order by the Prince Regent. In 1820 he was a founder member of the Astronomical Society of London, which became the Royal Astronomical Society in 1831. He and his wife Mary had one child, christened John, who was born at Observatory House in Slough on 7 March 1792. John Herschel had the best possible start in life for an astronomer, and he grew up to follow his father and become a leading figure in the astronomical world. While he was at Cambridge University John Herschel (1792–1871) befriended Charles Babbage (1791–1871), well known in computing circles as the designer of his difference engine, as well as the mathematician George Peacock

(1791–1858). These three started a movement to abolish what they called the 'dotty notation' of the differential calculus. The notation was widely used in England but it was not as logical as the dy/dx and d^2y/dx^2 terminology used by mathematicians on the Continent.

Between 1786 and 1802 William and Caroline Herschel compiled three catalogues giving a total of about 2500 positions of star clusters, nebulae and galaxies. Caroline's efforts as her brother's assistant were recognized by the crown, and in 1787 she was awarded a royal pension of £50 a year. She lived until 1848 and she was only a few days short of her 98th birthday when she died.

When John Herschel became involved with astronomical work one of his major tasks was the re-observation of the double stars already catalogued by his father. Using the large telescopes he was able to detect movements of some of these pairs of stars as they rotated about each other, and this was the first positive proof that Newton's law of gravitation was still valid outside the solar system. John Herschel was fortunate to find a competent collaborator called James South (1785–1867) who was wealthy enough to afford the refined instruments needed for this kind of work. The catalogue compiled between 1821 and 1823 by John Herschel and James South was published by the Royal Society in 1824, and it earned them both the Gold Medal of the Royal

Astronomical Society and the Lalande Prize from the Paris
Academy of Sciences.

Studying the Southern Heavens

In 1829 John Herschel married Margaret Stewart, who
was to bear him a large family of three sons and nine
daughters. He was very keen to complete his father's work
in astronomy and he knew that his father's researches
were confined to the Northern Hemisphere. He therefore
decided to undertake a journey south of the equator to
survey the skies not visible in England. He began plan-
ning his expedition in 1832 whilst he was still living at
Observatory House in Slough. He took his wife and children with him, and in November of that year John and
his family set sail for South Africa, taking with them a
large reflecting telescope for observing faint nebulae. They
also carried an extensive selection of astronomical instruments including a refracting telescope for observing double
stars. The Herschel family set up home and established
themselves in a farmhouse a short distance to the south
of Cape Town.

For four years John Herschel studied the clear skies
of the Southern Hemisphere and made excellent progress
during that time. By the time he and his family embarked
for home in March of 1838, John Herschel had recorded
the locations of thousands of stars and in addition he had

amassed long catalogues of nebulae and double stars. He added 1700 new entries to his father's catalogue, making a total number of 68,948 stars in the Herschel catalogues. After his return to England John Herschel was given the post of Master of the Mint – after Isaac Newton had held this post it was not uncommon for it to be offered to a scientist. John Herschel did not enjoy the pressures of the job at the Royal Mint, however, and he suffered a nervous breakdown in 1854.

13

UNDERSTANDING THE FORCES OF NATURE

The 19th century was a time of yet more crucial advances in the science of astronomy, marked by important events such as the discovery of another planet in the solar system. However, it was also a period of intense exploration, discovery and even controversy in the fields of physics, chemistry and geology – advances in all of which were critical for the progress of astronomy and cosmology in the decades to come.

In 1845 John Couch Adams (1819–92), a British mathematician and astronomer, had been carefully studying the irregularities in the motion of the planet Uranus. He correctly deduced that the deviations from the elliptical orbit were caused by another planet beyond Uranus, and he was able to calculate in which part of the sky the new planet could be found. He gave his data to James Challis (1803–82), director of the Cambridge Observatory, but although the data was accurate the Cambridge astronomers were unable to find the planet. The following year the French astronomer Urbain-Jean-Joseph Le Verrier (1811–77) made a very similar calculation and he, too, predicted that there was another planet beyond the orbit of Uranus. In September 1846 Le Verrier passed his information to the Berlin astronomer Johann Gottfried Galle (1812–1910). Galle and his assistant Heinrich Louis d'Arrest (1822–75) who painstakingly constructed a star map of the part of the sky suggested by Le Verrier, and in so doing they identified what they thought to be an uncharted star. The next night the 'star' had moved relative to the background stars. It was not a star at all, but the planet we now know as Neptune. Thus the British were narrowly beaten in their search to find and identify the next new planet, but on this occasion the French, who had so often lost narrowly to the British, certainly deserved their success.

Other Forces at Work

At this time there was much discussion about the nature of gravity. It seemed obvious that the force of gravity was one of the most important factors in astronomy and it was this force, acting over great distances, which determined the motions of the heavens. It was obvious that light crossed the distances between the stars, but gravity also managed to cross the distances between the planets. Moreover, according to Newton, every particle of matter in the universe attracted every other particle of matter, however small the attraction and however far apart they were from each other. Gravity was not the only force in the universe, however. In the 19th century scientists knew of two other forces, both of which had been known in one form or another for centuries. One of these was magnetism – the same force that enables a magnet to attract metal objects. The Earth itself has a magnetic field. It was easily shown that the whole Earth behaved like a giant magnet, and the north-seeking pole of the compass needle had been used for centuries to find direction at sea. It was of great value to navigation. The other force was discovered by the developing science of electrostatics. When particular substances like glass and amber were rubbed with certain kinds of cloth they acquired an electric charge. Rods charged in this way were able to attract small particles that clung to them.

Gravity seemed to be a very weak force compared with magnetism. It was possible to measure the gravitational effect of a mountain, but generally speaking it needed an object the size of the Earth to create a sizeable gravitational force. Nevertheless, gravity kept the planets in their orbits, and it seemed obvious that it was of more importance than the other forces when it came to the science of astronomy. Magnetism, on the other hand, did not need anything as large as a planet to produce a significant effect; it could be produced in quite small objects. It also differed from gravity in one other very important aspect. A north magnetic pole could not exist without a south magnetic pole. Opposite poles attracted each other but similar poles repelled. Whilst it was easy to demonstrate the existence of these opposing magnetic forces, no similar effect had been detected with gravity.

The electric force differed from both the gravitational and the magnetic forces. It appeared as both positive and negative charges and, as with magnetism, opposite charges were attracted to each other but like charges were repelled. Where the magnetic poles appeared in pairs of north and south, the electrostatic charges could be created in isolation without their counterpart. By the early part of the 19th century, scientists thought that electricity and magnetism must be related to each other, although the exact nature of the relationship was not known.

Advances in Magnetism and Electricity

Two British scientists are given much of the credit for advancing our knowledge of electricity and magnetism. One was Michael Faraday (1791–1867), who was born at Newington in Kent, the son of a local blacksmith. Faraday was a self-educated man who in 1812 became the laboratory assistant to Sir Humphrey Davy (1778–1829) at the Royal Institution in London. Faraday was the embodiment of the practical genius; his father taught him a lot about working metals, and he was an excellent laboratory assistant and a very skilled craftsman. Moreover, he was much more than a technician. He formulated theories of his own and he contrived experiments to prove his theories. He knew that magnetism could be used to create electricity, and he also knew that electricity could produce magnetism. He put his knowledge to practical use when he built a machine to generate electricity from rotary motion. It was the first electrical generator.

Faraday was not an astronomer, but it was through his work that James Clerk Maxwell (1831–79), the Scottish physicist, was able to discover the close relationship between magnetism and electricity. Maxwell formulated a set of equations relating the magnetic and electric fields, and towards the end of the 19th century his work became very valuable to astronomy. The equations showed that where it had been assumed that electrical and magnetic forces were different, they were in fact the same force – one could not exist without the other. To the astronomer, one important aspect of Maxwell's work was that his equations predicted the existence of other wavelengths of radiation, all travelling at the speed of light. Maxwell was working inside the electromagnetic spectrum at wavelengths in the microwave region, yet his equations held for radiation of any wavelength. Visible light had a wavelength of between 10^{-7} and 10^{-6} metres. Radio waves, which were not discovered until after Maxwell's time, have a wavelength ranging from about 1 metre to 10^5 metres. X-rays were discovered soon after radio waves, and they have wavelengths in the range 10^{-9} to 10^{-2} metres. Microwaves were then discovered, and they have wavelengths between 10^{-3} and 10^{-1} metres. The electromagnetic spectrum covered an incredibly wide range of wavelengths. But all forms of radiation, regardless of their different wavelengths and corresponding frequencies, always travelled at the speed of light and obeyed the equations of James Clerk Maxwell. Until the middle of the 20th century, however, any form of radiation outside the optical spectrum meant very little to astronomers. There was also a major problem with terrestrial-based observations, and this was the Earth's atmosphere itself. Our eyes had evolved to become sensitive to light in the visible part of the spectrum precisely because these were the wavelengths that the atmosphere allowed to filter through. Radiation at other wavelengths was invisible to our eyes, but even if we had evolved organs to see the radiation it would still not be visible on the Earth's surface because the atmosphere blocked it out.

Atoms and Elements

Great advances were also made in the field of chemistry during the 19th century. One of the pioneers was John Dalton (1766–1844), who was born in Cumberland and who spent much of his working life in Manchester. He revived the ancient Greek concept of atoms, but he brought the idea up to date and he developed it to become an exact science. Thus the idea of the atom was formulated in chemistry long before the physicists came to study it. By use of careful measurements Dalton was able to calculate ratios of the masses of the atoms of the elements. He studied many chemical reactions and he calculated the relative masses of all the components. He was able to create simple compounds such as carbon dioxide directly from the elements of carbon and oxygen, and he began the process whereby all chemical compounds became known by a formula giving the ratio of the elements they were made from. For example water was H_2O, carbon dioxide was CO_2 and methane was CH_4. His book, *A New System of Chemical Philosophy*, published in 1808, was a landmark in science and well ahead of its time. The idea of elements and compounds became well established and so, too, were the atomic weights of the various elements.

Dalton identified only 20 distinct elements in his early work, but more elements were discovered as his ideas continued to develop. By the middle of the 19th century, when the Russian scientist Dmitry Mendeleyev (1834–1907) came to study the subject, 63 elements had been identified. Mendeleyev devised a system in which he wrote the name of each element on a card and then tried to group the cards together to identify elements with similar chemical properties. He produced the earliest periodic table of the elements. In the periodic table, the known elements were put into groups that reflected their properties, similarities and differences. Sometimes Mendeleyev discovered gaps in the periodic table – this was an indication that an element was missing. In each case the missing element was eventually discovered.

It became known from experiment that every element had a set of spectral lines associated with it. These spectra were first determined in the laboratory by the simple method of burning the chemicals on a platinum wire using a Bunsen burner and focusing the light to pass through a triangular prism onto a screen. It became evident to the astronomers that very similar sets of spectral lines could also be seen in the Sun and the stars. This was proof that the stars were composed of the same elements as the ones on Earth. The most abundant element in the Sun was hydrogen, and the spectrum of hydrogen in particular proved to be of great importance.

Measuring the Universe

There were other important developments in astronomy about this time, and in the 1830s and 1840s came new evidence for the scale of the universe. Two different units evolved with which to measure the distance to the stars. One was the light year: this was the distance travelled by light in a year. The other unit was the parsec: this was the term formulated by the first astronomers to measure stellar parallax. The parsec is the distance to an imaginary star from which the maximum angular separation between the Earth and the Sun would appear to be 1 second of arc. A parsec is about 3.26 light years. It is a kind of reciprocal measure, the greater the number of parsecs the smaller the angle to be measured. For example, the Earth/Sun radius subtends an angle to 1 second of arc for a star at a distance of 1 parsec, but an angle of only 1/10th of a second of arc (not 10 seconds of arc) for a star at a distance of 10 parsecs. In 1838 the German astronomer Friedrich W. Bessel (1784–1846) made a direct measurement of the distance from the Earth to the star 61-Cygni, which he calculated to be situated at 3 parsecs. This is equivalent to a distance of 11.4 light years.

Relatively few stars in the night sky are close enough to be measured by the method of parallax. Fortunately for astronomers other methods of estimating distance became available in the 20th century, but all the new methods have to be calibrated by the stellar parallax method. The more stars that could be measured by parallax, then the more accurate the other methods became.

Finding the Age of the Earth

At this time few astronomers thought that the Earth was as old as the universe. However, the age of the Earth was still an important factor in the science of astronomy, as was the age of the Sun. It did not need a study of the skies to work out the age of the Earth; the rocks themselves were readily on hand for our examination. The biblical date for the creation, worked out from the genealogies in the Bible, was a mere few thousand years. According to theologians, the creation was calculated to have taken place in 4004 BC. In the 19th century this date was already under attack from archaeologists who discovered that some of the artefacts of ancient Egypt appeared to predate the creation. By the middle of the 19th century the literal interpretation of the Bible was being undermined even more, and it was seriously challenged by the theories of Charles Darwin (1809–82), formulated in the 1840s but remaining unpublished for 30 years. If Darwin's theories were true, it would have needed millions of years to allow for the evolution of species and for the development of mankind. There were other fields where long timescales were needed to explain their occurrence. The fossil record, for example, supported Darwin's theory of evolution and it, too, required an Earth timescale of many millions of years. The geological record pointed to an Earth that was hundreds of millions of years old. There was plenty of evidence to support the idea of a very ancient Earth.

There was one factor, however, that seemed to support the biblical account and to restrict the age of the Earth to a few thousand years, and this was the age of the Sun. It was not difficult to work out the immense amount of energy produced every second by the Sun. It did not matter how efficient the Sun was at producing light. Nor did it matter what combustible materials it had in its core. There seemed no way it could continue to burn at its current rate for more than a few thousand years without exhausting all its fuel. Life on Earth was not possible without the light and heat from the Sun. Indeed, Lord Kelvin (1824–1907) had 'proved' that the Sun would not be able to shine for more than a few thousand years – but of course he was basing his calculations on 19th-century science; he knew nothing of how energy is produced in the Sun by the process of nuclear fusion.

The British geologist Charles Lyell (1797–1875) was amongst the first to study the geological formations of the Earth and to try to establish dates for the geological record. From his travels in France and Italy he was able to observe a great many varied geological features and to work out how long it took for the sedimentary layers to form. In several places he discovered fossils of sea creatures high in the mountains, and this convinced him that the Earth's crust was moving very slowly but continuously all the time. Lyell's *Principles of Geology* was published in three volumes from 1830 to 1833. His system for determining dates for the geological record was based on two propositions. One was that the processes of geological change included all the causes that have acted from the earliest time. The other axiom was that these causes have always operated at the same average levels of energy. These two propositions suggest a 'steady-state' theory of the Earth. Changes in climate existed but they fluctuated around a mean, and they reflected the changes in the position of land and sea. Lyell was able to put the various rocks into a chronological order in terms of their formation, and he became convinced that the age of the Earth was measured not in terms of a few thousand years but on a timescale of hundreds of millions of years.

Astrophotography

The technique of taking photographs of astronomical objects like the Moon, the planets, the stars and all the other bodies in space is known as astrophotography. As photographic techniques developed, astronomers were quick to realize that the camera could be of great benefit to them in their studies of the heavens. English-born American scientist John Draper (1811–82) took the first photographs of the Moon in 1840. They showed lunar features such as the craters and 'seas'. His son, Henry (1837–82), also took pictures of the Moon and was the first person to create an image of a stellar nebula when he photographed the Orion Nebula in 1880.

The first photographs taken of the stars were disappointing, however: all that could be seen of the image were pinpoints of light. But photography proved to be very useful for the measurement of stellar parallax. In this technique bright stars were photographed against their dimmer and most distant companions. Then the photograph was repeated six months later using the same telescope and camera. The Earth had moved to the opposite end of its orbit during this period, and the nearer stars had therefore appeared to move by a small amount against their background. This tiny parallax was sufficient to make an estimate of how far away they were from the Earth.

As the method developed, an even better arrangement was devised. Astronomers flashed each image alternately onto a screen. Any star that had moved during the time the two exposures were taken could be identified immediately by its flicker. The star had a measurable parallax on the photographic plates and thus its distance could easily be calculated.

Today, astrophotography is an essential part of astronomy, and equipment such as the Hubble Space Telescope has revolutionized our understanding of space.

The Spectra of the Stars

As photographic techniques improved astronomers soon had other instruments at their disposal to provide even more information. A good example is the spectrograph. Stars had long been categorized by colour and brightness, but these were no longer the only criteria by which to measure the light from a star. The optical spectrum showed the light emitted at different frequencies, with well-defined spectral lines corresponding to hydrogen and other elements, and it enabled the stars to be classified by their different spectral types. Spectroscopy became one of the most useful tools to the astronomer.

Spectroscopy made great strides in the second half of the 19th century. In the 1880s Edward Pickering (1846–1919), working at Harvard College, established the Harvard photometry catalogue. This was the first catalogue to classify stars by photographic techniques. In Italy, a Jesuit priest called Angelo Secchi (1818–78) identified four distinct spectral types of star, and he became the first to classify the stars by their spectral properties. He was expelled from Italy by the revolution in 1848 and spent a short time at Stonyhurst College in England before moving on to Georgetown University in Washington DC. The Italian government recognized his work, and they allowed him to return to Rome in 1849 where he became professor of astronomy and director of the observatory at the Roman College.

Spanning the Ether

The nature of light was one of the many conundrums puzzling physicists. Did it consist of a stream of small particles, or was it a vibration in a medium that filled the whole of space? The particle theory to explain the transmission of light still prevailed, but opinion was swinging towards a theory based on wave motion. Wave motions were well understood; the sea could carry waves of widely different wavelengths and the air could carry sound waves with a wide range of frequencies. The problem for the astronomer, in attempting to understand how light travelled through space, was that waves needed a medium in which to travel. Thus, if light was a wave motion it followed that space must be filled with some kind of medium that could carry the vibrations of the wave, just as the air carried waves of sound. The medium was given a name; it was known as the ether. The ether was assumed to exist everywhere in the universe. When a ship travelled over the sea it had a velocity with respect to the sea. When a bird flew through the air it, too, had a velocity relative to the air. When the Earth travelled through the ether then it must have a velocity with respect to the ether. Scientists wanted to measure this velocity and thereby gain knowledge of the ether.

The American scientist Albert Michelson (1852–1931) devised an ingenious experiment to detect the motion of the Earth through the ether. A monochromatic light source was split into two beams. The beams were positioned at right angles to each other and, using mirrors, they were both reflected and brought together again to focus on an eyepiece where they produced a set of interference fringes. (The physics of interference fringes was well known and they were easy to detect.) The whole apparatus was mounted on a table that could be rotated so that the experiment could be performed at any inclination to the ether 'wind'. The theory was that one beam of light travelled in the direction of the ether wind and the other travelled transversely. As the apparatus rotated through a right angle it was predicted that the interference fringes should shift to the left or right by a fraction of a fringe.

The first experiment seemed to show a slight movement of the fringes, but after allowing for the errors of the observation the movement did not seem to be significant. Michelson, with his colleague Edward Morley (1838–1923), built a more delicate apparatus and repeated the experiment. Although the expected movement was only a fraction of a fringe, the apparatus was sensitive enough to detect a movement of one hundredth of a fringe. The experiment seemed infallible; it must be capable of detecting the motion of the Earth through the ether. But the experiment was repeated again and again and there was still no significant movement of the fringes. The scientists were puzzled. They could come to only one conclusion – either that the Earth was always travelling with the ether or perhaps it carried the ether with it as it orbited the Sun. This was a very unsatisfactory explanation, and as the scientific community argued about the findings some came forward with other suggestions. Eventually, the conclusion was reached that there was no ether, and so there must be another way for light to travel through space.

The Michelson–Morley experiment seemed to fail. It was unable to detect any sign of the ether. There was no medium to carry the waves of light. This problem was solved a few years later when the photon was discovered, showing that light was a stream of particles and not a wave motion. But physicists had to explain why the two light paths were equal in length when there should have been a very small measurable difference. Then the Irish physicist George FitzGerald (1851–1901) and the Dutch scientist Hendrick Lorentz (1853–1928) both independently suggested the same solution: the length of one arm of the apparatus had shrunk in the direction of motion. It defied logic, but they asserted that a body in motion, for example, a train travelling on a straight track, was seen by a stationary observer as shorter at speed than at rest. At normal speeds the contraction was only a few atoms, but at half light-speed the train would shrink to about 87 per cent of its stationary length. It all pointed to something very strange in the fabric of space/time and provided evidence for Einstein's special theory of relativity.

14

ALBERT EINSTEIN
Relativity Redefines Astronomy

One of the greatest contributors to modern physics, Albert Einstein (1879–1955) revolutionized the way scientists think about time, space, motion and gravity. He proposed the theory of relativity, reasoning that because the speed of light is constant, then distance and time, both of which define the speed of light, must be relative. His work also showed that it was possible to make an atomic bomb.

Albert Einstein was born in 1879 at Ulm in Germany. The following year his family moved to Munich. In 1895 they moved to Zurich, where the 16-year-old Albert failed his entrance examination in engineering to the Zurich Polytechnic. His early education was undistinguished, and years later he wrote with a lack of enthusiasm about the strict teaching he endured at the Luitpold Gymnasium when he was about 15 years of age:

When I was in the seventh grade at the Luitpold Gymnasium I was summoned by my home-room teacher who expressed the wish that I leave the school. To my remark that I had done nothing amiss he replied only 'your mere presence spoils the respect of the class for me.' I myself, to be sure, wanted to leave school and follow my parents to Italy. But the main reason for me was the dull, mechanised method of teaching. Because of my poor memory for words, this presented me with great difficulties that it seemed senseless for me to overcome. I preferred, therefore, to endure all sorts of punishments rather than to learn to gabble by rote.

An Unpromising Start

Albin Herzog, the director of the Zurich Polytechnic, urged Einstein not to give up his endeavours, but to seek entrance to the progressive Swiss Cantonal School in nearby Aargau. Einstein followed the advice, and the school turned out to be the making of him. He lodged with one of the teachers, Jost Winteler, where he was treated as one of the family. In fact he became one of the family, for one of the Winteler's sons married Einstein's sister Maja. One of their daughters even married his friend Michele Besso.

After completing his schooling Einstein became a part-time teacher and a private tutor until 1902, when he was offered a job at the Swiss Patent Office in Bern on the modest salary of 3500 francs a year. At the age of 23 he had still achieved nothing in the academic world, but at least he had a regular job, and also one that wasn't too onerous – it gave him plenty of time to speculate about questions of philosophy and the nature of the universe. Einstein did not conform to the popular image of the quiet cloistered academic, however. At Bern he made two close friends, Konrad Habicht (1886–1958) and Maurice Solovine (1875–1958), and the three called themselves the 'Olympian Academy'. They discussed philosophy, literature and physics in a noisy and boisterous fashion well into the night, much to the annoyance of their immediate neighbours.

In 1903 Einstein married Mileva Maric. The union was blessed with a son called Hans Albert, born in 1904, and by a second son, Eduard, born in 1910. Albert Einstein's academic career was still progressing very slowly at this time, but in 1905, at the age of 26, he submitted a PhD thesis to the University of Zurich. The thesis was rejected as being too short. Einstein put in one extra sentence and resubmitted the thesis. This time it was accepted.

Some Relative Thoughts

It was at Bern, whilst he was travelling along the high street by tram, that Einstein first began to wonder about relative motion. If the tram were to speed up to approach the velocity of light, then he realized that the time shown on the town hall clock ahead of him would appear to speed up because the light from it had a progressively shorter distance to travel. He surmised that on his return journey, when the tram was speeding away from the clock, the clock would appear to slow down, for the light took progressively longer to reach him. If the tram reached the speed of light then the clock would remain at a fixed time, he surmised – in other words, time shown on the clock would appear to stand still. He calculated the equations relating space and time in the two frames – one frame being the high street in Bern with the clock, and the other being the moving tram – and formulated what he called the principle of relativity: that the laws of physics were the same in all uniformly moving frames. He showed how time and distance could be different as seen by two or more observers in different frames, but the laws of physics were the same.

Einstein pondered the ideas of relativity for several years before giving his first findings to the world. His first published paper, in 1905, was a description of the phenomenon known as the photoelectric effect. This was a property observed in certain metals. When light shone on the metal, electrons were emitted. Einstein explained the effect in terms of particles of light striking the atoms of the metal and causing them to release electrons. This theory alone would have won Einstein fame, but it was followed very quickly by something much more radical. It was his account of the principle of relativity – now called the special theory of relativity – in which he expounded his ideas about moving and fixed frames. His famous imaginary experiment involves a frame moving at constant speed relative to a fixed frame, such as a train moving along a straight track. Now, imagine a stationary observer situated by the side of the track halfway between two signal lights. Both the lights flash at the same time, and the observer, being positioned an equal distance away from both, concludes that the lights were switched on at exactly the same time.

Next, imagine another observer travelling on a train between the two simultaneously flashing lights. The observer on the train also sees the lights flash, but he is moving towards one light and away from the other. So he sees the light he is travelling towards before he sees the light he is travelling away from because it has less distance to travel. To this observer, the lights flash at different times. According to Einstein's theory of relativity, different people do not therefore necessarily see the same event at the same time.

Albert Einstein believed that the speed of light was exactly the same in all frames of reference. He believed that if the observer on the train were to measure the speed of the light from the two stationary signals then he would arrive at the same result as the ground-based observer. This suggestion, taken to its ultimate, leads to some astonishing conclusions. In the moving frame of the train, for example, all lengths in the direction of motion are seen as foreshortened from the rest frame of the Earth. But according to Einstein's theory there was really no such thing as a privileged rest frame (in other words, a frame that does not itself move while all other things are moving relative to it); every observer in a uniformly moving frame could assume that he or she was stationary whilst the rest of the world moved past. Space and time were related in such a fashion that time appeared as a fourth dimension. Einstein's theory that light always travelled at the same speed was difficult to prove, but in fact the proof was already there. It was the conclusion reached by Michelson and Morley in their seemingly negative experiment conducted in the previous century.

Some of the conclusions of Einstein's special theory of relativity were to have repercussions in the field of astronomy. Using Einstein's hypothesis, the mass of a body increases with its speed. The closer the speed approaches to that of the speed of light the heavier the body becomes, so that if it can be accelerated to reach light speed then its mass will become infinite. It follows, therefore, that it is impossible for anything to travel faster than light.

This was a serious blow for astronomers. It meant, for example, that within a human lifetime it would be impossible to send a message to star systems and to get a message back again, except in the case of a few close systems. It also meant that the universe would be much more difficult to explore, and for humans to travel through, than was first thought.

Laboratories in the Mind

As soon as Einstein's paper was published he was formulating another theory destined to be just as earth-shattering. It concerned the relationship between acceleration and gravity. He was struck by the fact that these seemingly very different concepts produced some very similar results. In another thought experiment, Einstein envisaged an earthbound laboratory. In it, the experimenter could verify the laws of mechanics and conclude that all matter was drawn to the Earth by the force of gravity. In the laboratory, a body would fall with an acceleration g, just as 300 years earlier Galileo's musket and cannon balls had fallen from the leaning tower of Pisa in his experiment on the study of gravity.

Einstein then imagined a second laboratory that was effectively a spacecraft. This craft was equipped with engines that could propel the laboratory through space with a constant acceleration of g. The experimenter in this imaginary craft could also stand in his laboratory and demonstrate that any object falling to the floor appeared to fall with an acceleration of g.

The gravitational field of the earthbound laboratory and the acceleration of the other laboratory in space gave exactly the same results for the experiments on the laws of mechanics. Then Einstein formulated what he called the principle of equivalence. He claimed that if his experimenters could not perceive the world outside their laboratories then there was no way of telling the difference between the gravitational field and the constant acceleration. The effects were equivalent.

Einstein went further. He maintained that the sceptical scientist could use something to devise an experiment to tell which type of laboratory he was in. The 'something' was a beam of light. Light travelled in a straight line. Therefore if a beam of light crossed the earthbound laboratory the observer would measure its course as a straight line. If the observer in the accelerated space laboratory performed the same experiment, however, then the acceleration would cause the light to appear to be slightly deflected down to the floor of the laboratory.

Imagine the light beam entering horizontally on the left-hand side of the laboratory. By the time the beam has crossed to the right-hand side the accelerating laboratory has moved, thus the light beam appears lower; its path is bent towards the floor of the laboratory. The sceptical scientist did not know that gravity could bend the light so he assumed that he had found a difference between the two laboratories.

But Einstein insisted that there was no measurable difference between the two laboratories. He explained his assertion by claiming that in the earthbound laboratory gravity would draw the light beam to the floor by exactly the same amount as measured in the accelerated laboratory. One of the basic assumptions in Einstein's new theory was that a beam of light was deflected by a gravitational field. The deflection was very small and there was no way to measure it in an earthbound laboratory. But it could be measured during a total eclipse of the Sun. If there were stars near the rim of the Sun during an eclipse, then the light from those stars would be deflected by the Sun's gravity and the stars would appear to be pushed outwards to different positions for the duration of the eclipse.

Proved by an Eclipse

In 1911 Einstein calculated that the deflection of the starlight by the Sun would be less than 1 second of arc –

in fact, he calculated it to be 0.83 seconds, a tiny deflection but measurable with the techniques of the day. Unfortunately, World War I (1914–18) prevented Einstein from proving his theory at the next suitable eclipse. Einstein did not let this misfortune halt his research, however, and he sought other evidence to support his theories. He knew that in the 19th century the astronomer Urbain-Jean-Joseph Le Verrier (1811–77), who predicted the position of the planet Neptune, had also made some careful measurements of the planet Mercury. Le Verrier was able to show that the orbit of Mercury was not a simple ellipse, but one that precessed very slowly around the Sun. It was a triumph of astronomy to measure the amount of the precession – only 38 seconds of arc in a whole century. Le Verrier tried to explain the precession in terms of an unknown planet closer to the Sun than Mercury. An amateur astronomer claimed to have discovered the planet, and it was given the name Vulcan – but in fact Vulcan did not exist and the precession still remained unexplained.

Einstein knew of Le Verrier's result and he realized that the precession could be explained by the theory of relativity – the relative bending of space under the influence of gravity. He calculated a figure of 43 seconds of arc per century for the advance of the perihelion. It was a brilliant result and was highly acclaimed, but it raised a few eyebrows. After all, the precession of Mercury had been discovered long ago by Le Verrier, and Einstein therefore already knew the answer to the question of how relativity affected the orbit of Mercury. Had he predicted the precession before the findings of Le Verrier had been known, then that would have been far more impressive. Einstein knew that the next suitable eclipse of the Sun would prove his assertions.

However, he had to wait a few more years during which time he revised his calculations. He decided his figures were out by a factor of two, and his new prediction for the bending of light by the Sun was 1.7 seconds of arc. When World War I was over, two British expeditions were sent to observe an eclipse of the Sun. One went to Sobral in Brazil and the other headed for the tiny Portuguese island of Principe off the west coast of Africa. On 23 May 1919 the sky darkened at those two places just as the astronomers had predicted. In Principe the rain came just as the eclipse was due, but observer Arthur Eddington (1882–1944) was still able to take his photographs, and he measured the star positions with enormous excitement. In Brazil the plates were also measured with great care. Both parties found that the light from the stars near the rim of the Sun were deflected by exactly the amount that Einstein had predicted earlier.

It was a great moment and the astronomers could not contain their joy. Einstein's principle of equivalence had been vindicated. It opened up a new and far-reaching theory of relativity that became known as general relativity. Einstein was able to show that space consisted of a curved space–time medium in which the stars and the planets bent the fabric of space–time with their gravitational fields. It was as a result of general relativity that Einstein discovered a relationship between matter and energy, expressed in terms of the simple equation $E = mc^2$.

'The Biggest Blunder' of his Life – the Cosmological Constant

At the time Einstein developed general relativity, the universe was thought to be static, with only random motions of other galaxies towards and away from our galaxy. His work, however, predicted the inexorable contraction of the universe due to all the galaxies exerting a gravitational pull. Einstein thus found it necessary to modify his field equations by the inclusion of a 'cosmological constant' that represented an outwards pressure associated with otherwise empty space. Such a force could act as a counterbalance against gravity to produce a stable, eternally unchanging universe.

Not all scientists were convinced of the need for the cosmological constant in Einstein's equations. Alexander Friedmann (1888–1925), a Russian meteorologist and mathematician, dispensed with the cosmological constant to investigate solutions where the universe could change and expand with time. He proposed the radical idea of an evolving universe in an article in 1922, but the result was largely ignored by the establishment, and Friedmann received no recognition of his idea. Subsequently Georges Lemaitre (1894–1966), a Belgian astronomer and priest, independently experimented with the value of the cosmological constant to conclude that the universe was expanding, just two years before Edwin Hubble (1889–1953) published his correlation between the distance and velocity of galaxies. Lemaitre also extended the idea of a changing universe, by following the model through time in reverse to infer a single point of creation. This is the first suggestion of an initial 'Big Bang' moment.

The work of both Friedmann and Lemaitre was strongly criticized as irrelevant by Einstein at the time, although he quickly recanted after Hubble's discovery and thereafter publicly supported Lemaitre's interpretation. Einstein also then discarded the cosmological constant, allegedly dismissing it as 'the biggest blunder' of his life.

Einstein and Gravitational Lensing

A gravitational lens is formed when the light from a very distant, bright source (such as a quasar) is 'bent' around a massive object (such as a cluster of galaxies or black hole) between the source object and the observer. The process is known as gravitational lensing, and is one of the predictions of Albert Einstein's general theory of relativity.

Initially Einstein only considered a form of gravitational lensing by single stars (as confirmed by Sir Arthur Eddington in 1919 during a solar eclipse). Although the Russian physicist Orest Chwolson (1852–1934) is credited as being the first to discuss gravitational lensing in print (in 1924), the effect is more usually associated with Einstein, who published a more famous article on the subject in 1936. In 1937 the American-based Swiss astronomer Fritz Zwicky (1898–1974) first considered the case where a galaxy could act as a lens, something that according to his calculations should be well within the reach of observations. However, it was not until 1979 that the first gravitational lens would be discovered. It became known as the 'Twin Quasar' since it initially looked like two identical quasars; it is officially named Q0957+561.

Gravitational lenses act equally on all kinds of electromagnetic radiation, not just visible light. The phenomenon often creates streaks and arcs out of the lensed object. If the source, massive lensing object and observer lie in a straight line, the source will appear as a ring, which is often referred to as an Einstein Ring. If the 'lens' is very symmetrical, then the source appears as four images symmetrically arranged around the lensing object. This is known as an Einstein Cross, or Huchra Lens, after the American astrophysicist John Huchra (born 1948), who first discovered it in 1985.

A Safe Refuge

Einstein had spent a great deal of time on his theory but, like Isaac Newton before him, he found that he lacked the mathematical tools to formulate his ideas and develop them. He was able to turn to his staunch friend Marcel Grossmann (1878–1936) in his time of need. Grossmann knew about tensor calculus (used to deal with the mathematics of four dimensions) and the curvature of space. Using the shorthand notation of tensor calculus a great deal could be expressed with relatively few equations, but the field equations were formidable and solutions could only be found in a few special cases. As more and more mathematicians became involved other solutions to the equations were found. But Einstein was never satisfied with his efforts. James Clerk Maxwell had shown that the electric and magnetic fields were one and the same. Einstein wanted to unify the electromagnetic field and the gravitational field into a single, unified field theory.

In the 1920s there came many new and exciting advances in physics. Einstein attended all the important conferences in these years and he was well aware of the developments in quantum mechanics. He could not believe that the physics of the atom depended upon probabilities and uncertainties. His famous saying was that *'God did not play dice with the universe.'* But in this respect Einstein's instinct was proved wrong and because of this he became isolated from many of the scientists of his time.

In the 1930s, as Hitler and the Nazis rose to power, Einstein was able to foresee all too clearly the future development of Germany. He was the world's most famous scientist by this time and he had plenty of contacts in England, in other European countries and in the United States of America. He had no problem in finding a sinecure well away from Germany, and in 1933 he left for America accompanied by his second wife Elsa, his secretary and his collaborator Professor Walter Mayer (1887–1948). He had a choice of practically any university in America in which to work, but he chose Princeton, New Jersey, as the place to spend his later years. Here, safe from persecution, Einstein was able to speak out freely against the Nazis.

A New and Deadly Dawn

During World War II (1939–45), the possibility of building atomic weapons came to the fore. There were two major questions to be answered. First, was it possible to build a weapon that would convert matter into energy as predicted by Einstein's famous equation $E = mc^2$? In other words, was it possible to build an atomic bomb? Second, if it was possible to make such a bomb, then could the Nazis already be working on a similar project? The answers to both of the questions were not particularly surprising, and nor was the outcome. The US government wrote to Einstein asking him if it was indeed possible to manufacture an atomic bomb. He had no option but to reply in the affirmative, and the US government immediately set up a programme to build one. The end result was the terrible destruction of Hiroshima and Nagasaki by atomic bombs. At the end of the war it was discovered that the Nazis had no plans for making an atomic bomb, and it left Albert Einstein feeling for the rest of his life that he had been somehow responsible for the development of nuclear weapons.

In the 1920s the sciences of atomic physics and nuclear physics were both destined to play a major part in the advancement of astronomy. Einstein had redefined the laws of mechanics and astronomy. Newton's laws still held good for most terrestrial observations, however, and it was only when velocities comparable to the speed of light were involved that Einstein's theories took precedence. Einstein displayed the portraits of three of his most admired predecessors on the wall of his study. All three were British: they were Isaac Newton, Michael Faraday and James Clerk Maxwell. He felt some guilt at having destroyed the models of the universe so painstakingly put together by his predecessors. In later life, when he was working on his autobiographical notes, he found himself making a list of the difficulties he had discovered in the Newtonian system. He suddenly stopped himself in his tracks and he addressed Newton directly and movingly:

> *Enough of this. Newton forgive me. You found the only way that, in your day, was at all possible for a man of the highest powers of intellect and creativity. The concepts that you created still dominate the way we think in physics, although we now know that they must be replaced by others farther removed from the sphere of immediate experience if we want to try for more profound understanding of the way things are interrelated ...*

The last 20 years of Einstein's life were pleasant and peaceful. He still worked on his field equations and in 1950 he produced another set of solutions. They were never published, and validators soon discovered errors in his work. Despite this, he is remembered as the greatest scientist of the 20th century, even though he was never able to reach his final objective, to unify the forces of nature.

15

THE HUBBLE UNIVERSE

By the time Albert Einstein's ground-breaking theory of relativity was published in 1905, astronomy had made considerable progress in several other important areas. The development of new techniques and more powerful telescopes was beginning to enable astronomers to measure the distances of far-away stars and galaxies more accurately, indicating that the universe was much larger than anyone had previously imagined, and was expanding and evolving.

America's contribution to astronomy was minimal until the period after their civil war. The stability that followed the American Civil War (1861–5) allowed the US economy and the universities to expand rapidly, and more money became available for research. The contribution of the Americans then became very significant.

Measuring Brightness and Distance

Henrietta Leavitt (1868–1921) was born in Lancaster, Massachusetts. From 1886 to 1888 she attended the Society for the Collegiate Instruction of Women, later known as Radcliffe College, where she graduated in 1892. In 1895 she became a volunteer assistant at Harvard Observatory and in 1902 she was given a permanent staff appointment. The observatory's great astronomical project, started by Edward Pickering (1846–1919), was to measure the brightness of all the known stars as accurately as possible. Leavitt was appointed head of the photographic stellar photometry department, and with rapidly improving photographic techniques the observatory was soon measuring stellar magnitudes to a greater accuracy than anyone had achieved before.

A new phase of the work began at Harvard in 1907 when Edward Pickering set up an ambitious plan to ascertain the stellar magnitudes photographically. Unlike earlier measurements by eye, which could be subjective, the photographic plates gave a truer reading of the colours of the stars. Henrietta Leavitt began by studying a sequence of 46 stars in the vicinity of the North Celestial Pole. With the new methods of analysis now available she determined all the magnitudes, and then followed up her work with a much larger sample in the same region of the sky. She extended the scale of standard brightness from the

3rd magnitude right down to the 21st magnitude, with her work being published progressively between 1912 and 1917. By the time of her death, Henrietta Leavitt had determined the magnitudes for stars in 108 areas of the sky. She had also discovered four new stars and 2400 variable stars. In fact, during the 1920s, she had discovered more than half of all the known variable stars in the sky.

Leavitt's most valuable contribution to astronomy, however, was her discovery in 1912 of the properties of a certain class of stars known as the Cepheid variables. Cepheids are like lanterns flashing in the sky – the brightness of a Cepheid is not constant but oscillates over a period of time that varies from a few days to a few months. Leavitt's breakthrough came when she compared the relative brightness of Cepheid variable stars in the Small Magellanic Cloud, all of which could be assumed to lie at approximately the same distance from Earth. She discovered that the maximum absolute brightness of this class of star was closely related to the period of the luminosity variation (the regular changes from rapid brightening to gradual fading). Thus she was able to derive a simple formula to calculate the distance of any Cepheid by measuring its period and maximum luminosity. This was a major advance in our understanding of the scale of the universe, and it provided an alternative technique to the method of parallax for determining distances. More importantly, it enabled the distance to a Cepheid variable to be measured at much greater distances than had previously been possible by the parallax method.

The Spiral Nebulae

Observational astronomers were trying hard to understand the nature of the faint fuzzy clouds visible in the sky – such as the 'nebulae' that had been catalogued by Messier and Herschel. Some were clearly stellar in nature, such as the globular clusters. Others appeared as hazy clouds containing a few bright stars, which were eventually identified as star-formation regions within the galaxy. The most puzzling, however, were the small nebulae that showed a spiral structure, but without clearly resolved individual stars.

At the Lowell Observatory in Flagstaff, Arizona, Vesto Slipher (1874–1969) was using the 61-centimetre (24 in) refractor to make spectroscopic observations of the spiral-shaped nebulae. The nebulae were incredibly faint, and each measurement required very long exposures of many hours. The spectra showed a phenomenon known as 'redshift'. The light reaching Slipher's telescope from the nebulae invariably showed the familiar absorption lines, but shifted towards the red end of the spectrum.

There were two reasons for redshift to occur and

both were fairly well understood. One possibility was that the light was created in a very strong gravitational field. (Einstein predicted this effect, although he applied it to individual stars rather than to a whole galaxy.) The other explanation for redshift was if the emitting object was moving away from the Earth at a very great speed, analogous to the 'Doppler effect'. In 1842 Christian Doppler (1803–53) had shown how the frequency of sound changes if it is emitted by a source travelling either towards or away from an observer. This is the reason for the rise and drop in note of a racing car as it rushes past. The wavelength of light emitted by a moving object can similarly be shifted to appear either bluer or redder, according to the speed of the emitter relative to the observer. The shift in spectral features was already used in observational astronomy as a technique to assess the motions of nearby stars in the Milky Way.

In 1917 Slipher published the astonishing results of spectroscopy for a sample of 25 spiral nebulae. Only four of these were approaching the Earth, with the other 21 receding, giving an overall sense that the nebulae were mostly 'scattering' away from us. If due to the Doppler effect, the velocities implied by the redshift were hundreds of kilometres per second – far greater than any speeds observed for other known galactic objects. Slipher continued to observe a further 20 spiral nebulae, which were all found to be receding. However, the cause of the dispersion of the nebulae was to remain, for the moment, a mystery.

The 'Great Debate'

The nature of 'spiral nebulae' such as the Andromeda Galaxy was a subject of hot debate, with implications not only for the size and structure of the galaxy but also the size of our whole universe. Either the nebulae were intrinsically small structures contained within our Milky Way, which in itself comprised the entire universe, or they were intrinsically large objects that appeared small because they lay well outside our own galaxy.

The 'Great Debate' was held on 26 April 1920 at the Smithsonian Museum of Natural History in Washington DC, between the two eminent American astronomers Harlow Shapley (1885–1972) and Heber Curtis (1872–1942). Shapley assumed that our galaxy was large, and therefore for Andromeda to be the same size it would need to lie at an immense distance, thus rendering the entire universe unimaginably big. He thus proposed the spiral nebulae were located within our galaxy, which comprised the whole universe. On the other hand, Curtis maintained that the spiral nebulae were independent 'island universes' located outside our own galaxy. He favoured a model in which the Sun was (incorrectly) placed at the centre of a much smaller galaxy. In his support, Curtis pointed out that unlike known galactic objects, the spiral nebulae were not observed to lie in the band of the Milky Way, and they had discrepant redshifts. He also noted that they had dust lanes similar to those observed in our own galaxy, and that the Andromeda Nebula had produced many nova stars, which would be unusual if it were only part of our galaxy.

The issue was resolved only five years later, when Hubble published the first measurement of the distance to Andromeda, proving beyond doubt the extragalactic nature of the spiral nebulae. Despite his erroneous model of our galaxy, Curtis had been proved right.

Getting a Better View

In the 1920s California came to the forefront of astro-nomical research. The main reason for its prominence was the building of the Mount Wilson telescopes, situated high in the San Gabriel Mountains of southern California. These were two large telescopes which not only offered very clear views of the skies from their high altitude position but were also the largest telescopes in the world.

The Mount Wilson Observatory was founded in 1904 by astronomer George Ellery Hale (1868–1938). He became its first director, and was responsible for hiring both Harlow Shapley and Edwin Hubble. For many years the Mount Wilson Observatory and the neighbouring Palomar Observatory, near San Diego, were operated jointly as the Hale Observatories by the Carnegie Institution of Washington and by the California Institute of Technology in Pasadena. This arrangement continued until 1980 when the observatories became separate units.

The Mount Wilson Observatory holds a number of optical telescopes. The most significant of these are the 1.5-metre (60 in) reflector and the 2.5-metre (100 in) reflector. In the early part of the 20th century the 2.5-metre (100 in) reflector was the world's most powerful telescope, and with it Edwin Hubble and his associates made important discoveries about the distant galaxies, notably their extragalactic nature and motion away from the Earth. These results were the foundation of a new view of the universe, and of our galaxy's place within it.

In 1914 Harlow Shapley (1885–1972) joined the staff of the Mount Wilson Observatory. He used the observatory's 1.5-metre (60 in) reflecting telescope to study the distribution of globular star clusters in the Milky Way Galaxy. These clusters are dense groups of stars, each containing hundreds of thousands of members packed into a tight spherical ball. He found that of about 100 clusters known at that time, a third of them lay in the constellation of Sagittarius. In the star clusters Shapley was able to identify pulsating stars very like the Cepheid variables. In fact they were of a type now known as RR Lyrae variables. These have a shorter period than the Cepheids, but are the same class of star. Using Henrietta Leavitt's period-luminosity relationship, Shapley could therefore calculate the distance to 93 globular star clusters. He constructed a three-dimensional map of the stars centred on a point in Sagittarius that he assumed was the middle of the Milky Way. From this conclusion and his other distance data Shapley deduced that the Sun lay at a distance of 65,000 light years from the centre of the galaxy, although this distance was later modified to 30,000 light years. Before Shapley's work the Sun was believed to lie near the centre of the galaxy, but he not only proved this to be wrong, he also made the first realistic estimate for the actual size of the galaxy. It was another milestone in galactic astronomy.

The Universe's True Scale Revealed

Edwin Hubble gained a degree in mathematics and astronomy at the University of Chicago. Whilst there, he was inspired by the astronomer George E. Hale (1868–1938). Hubble was also known for his athletic prowess, and gained a reputation as a boxer. After graduating, he decided on a change of career path, and travelled to England as a Rhodes scholar to study law at Oxford University. In 1913, on his return to America, he joined the Kentucky bar but found himself bored with law and dissolved his practice soon afterwards. He returned to the University of Chicago and to the Yerkes Observatory at Wisconsin; whilst there he gained a PhD in astronomy. Hubble served in World War I (1914–18), and after the war he settled down to work at the Mount Wilson Observatory in California where he began to make his discoveries concerning the distant galaxies.

Hubble left Mount Wilson in 1942 hoping to join the armed forces again and to serve in World War II (1939–45), but he found that he had more to offer as a scientist. After the war he went back to Mount Wilson and was able to convince his employers that they should build an even greater telescope than the 2.5-metre (100 in) reflector. He was instrumental in the design of the Hale Telescope at Mount Palomar Observatory. Hubble died in 1953. In recognition of his work the first large orbiting space telescope, launched in 1990, was named after him. Today the Hubble Telescope is almost synonymous with spectacular photographic images of distant stars and galaxies.

It was while he was at the Mount Wilson Observatory that Hubble observed Cepheid variable stars contained within the Andromeda Nebula. Using Leavitt's relation between the period of variability and the brightness of

Cepheids, he derived a distance of some 900,000 light years to the nebula, so large that Andromeda clearly could not be part of our own galaxy. Similarly, his observations of the spiral object now known as the nearby Triangulum Galaxy showed that it, too, must lie beyond our Milky Way. The enormous distances also implied that the 'nebulae' that the astronomers observed as faint and tiny, were in reality incredibly bright and large. When Hubble made his results public in 1925, astronomers accepted the conclusion that the spiral objects could only be other galaxies outside our own. The findings also demonstrated for the first time the insignificance of our Sun, being one amidst thousands of billions of stars in a galaxy which in its turn is just one of billions.

Classification of Galaxies

Having established both the nature and mind-boggling distances to the other galaxies, Hubble set out to discover their properties, and to use them to find out more about the universe. He soon realized that galaxies had a variety of different shapes, and characterized these shapes in a classification scheme that spanned the range of elliptical, spiral, barred spiral and irregular galaxies. Within each class there were subdivisions based on the detailed structure: the degree of ellipticity in an elliptical galaxy; how tightly the spiral arms were wrapped, and the relative brightness of the bulge and disc in the spiral systems. The irregulars were more difficult to classify, and we now suspect they form in the gravitational chaos when two or more galaxies collide with each other. Hubble published this scheme as a 'tuning fork' diagram, which he thought represented an evolutionary sequence (an idea which has subsequently been disproved). He also estimated the mass of each type of galaxy, and made the first measurements of the matter density of the universe.

An Expanding Universe

But Hubble had another major discovery to make. He continued to assemble measurements of the distances to the galaxies, but was curious about Slipher's early results showing their almost uniform recession. This was completely contrary to the expectations of a static universe, which predicted a random but balanced distribution of galaxy motions both towards and away from the Earth. Working with his assistant Milton Humason (1891–1972), he first confirmed Slipher's spectroscopic results. It was when he combined these velocities with his distance estimates to the galaxies that an astounding result emerged. The galaxies described a clear proportionality between distance and velocity – galaxies five times as far away were moving five times as fast, and galaxies ten times further away moving ten times as fast.

In 1929 Hubble first published this result for the 20 galaxies for which he was most confident of the data, and two years later confirmed the relationship also held for galaxies beyond a distance of 100 million light years and at velocities up to 12,427 miles per second (20,000 km/s). This showed that light from the furthest galaxies had taken many million light years to reach the Earth. In other words, he was looking at such galaxies as they were 100 million years ago!

Hubble's findings posed many questions about the nature of the universe. The direct relationship between velocity and distance revealed a pattern in the motions of the galaxies that is a natural consequence of a systematic and constant expansion. This was the first experimental suggestion for an expanding universe; the few nearby galaxies showing a discrepant blueshift could be explained as those that were responding to the local gravitational attraction of the Milky Way. The distances to all other galaxies are increasing with time, and the overall mass density of the universe steadily decreasing. The observational evidence for an expansion paved the way for the concept of the Big Bang, an explosion of immense energy responsible for the creation of all of space, matter and energy at the beginning of time. If the galaxies were now expanding away from each other, then in the past they were closer together. Go back sufficiently far in time, and the galaxies would all be compressed into a much smaller region. The Milky Way does not have a 'special' place representing the centre of the expansion – all the galaxies are receding from one another, and Hubble's law could have been derived from observations of any galaxy.

The velocity–distance relation became known as 'Hubble's law', with the scaling between the two properties known as 'Hubble's constant', which is measured in units of km per second per Megaparsec (Mpc). (A Megaparsec is one million parsecs; a parsec is a distance equivalent to 3.26 light years.) Hubble's constant gives a measure of the present rate of expansion of the universe, and offers an estimate for its lifetime. For any galaxy travelling at constant speed over a known distance, the ratio of the two attributes allows an estimate of the time it has taken. Thus the inverse of Hubble's constant yields an estimate of the age of the universe, assuming expansion has remained uniform. The first value for Hubble's constant was 500 km/s/Mpc, from which the universe was deduced to have an age of two billion years. The accurate evaluation of Hubble's constant was an important aim throughout 20th-century astronomy, and for many decades it was in error by over 50 per cent. Its determination was one of the key projects guiding observations with the Hubble Space Telescope. It is only in the first years of the 21st century that measurements pinned down the Hubble constant to a value of 70 km/s/Mpc, revising the resulting estimate of the universe's age closer to 14 billion years.

Hubble's discovery did not just revolutionize how astronomers viewed the history of the universe, overturning the idea of a static universe in favour of one that was evolving with time, it also pushed them to consider its future over the next billions of years. If the universe were expanding, would it ever stop? There were three immediate possibilities, and which was correct would depend on the total mass density of the universe. The expansion of space was carrying the galaxies further and further from each other. If the total mass of the universe was high, then the combined gravity would eventually slow the expansion – at some instant the galaxies would cease moving, and then pull back towards each other in an inevitable 'Big Crunch' marking the end of the universe. At the other extreme, if the universe were comparatively empty, there would be insufficient gravity ever to draw the galaxies back together, and the expansion would continue until the galaxies moved so far apart that they would no longer be visible to each other. The third option was the boundary point between these two eventualities, with a critical mass density of the universe that would be enough to decelerate but not reverse the expansion, with the galaxy motions coming to rest only at infinity.

An Alternative to the Big Bang

Before Hubble's work suggested the universe started with the Big Bang, a theory for a static universe had been put forward by Sir James Jeans (1877–1946) in about 1920. It was developed after World War II by Fred Hoyle (1915–2001) at Cambridge University as a rival to the Big Bang theory; Hoyle called it the steady-state theory. He proposed a universe where matter was created from nothing. The production of only a few atoms per year would be sufficient to cause the universe to expand as observed by Hubble. The idea of creating matter from nothing did not appeal to many astronomers, but the steady-state advocates pointed out that the Big Bang theory required a whole universe to be created out of nothing as well. The steady-state theory did at least encourage discussion about alternatives to the Big Bang as a way of explaining the origins of the universe. Hermann Bondi (1919–2005) and Thomas Gold (1920–2004) raised it again in a revised form in the 1980s, but by then the evidence in favour of the Big Bang was almost conclusive.

16

FROM MICROCOSM TO MACROCOSM

To truly comprehend the nature of the stars and the galaxies it was necessary to develop areas of science such as nuclear physics and quantum mechanics. This new knowledge provided astronomers with an insight into the way stars create such vast amounts of energy for so long and, ultimately, how they die.

Early in the 20th century the word 'nebula' was applied to an object in the sky that had a nebulous or poorly defined appearance compared with the sharp points of light that corresponded to the main sequence stars. It soon became apparent that there existed several kinds of nebulae. One kind was a cloud of gas, such as could be seen in the Horsehead Nebula in the constellation Orion. The other kind was a distant galaxy that had, for example, nebulous spiral arms. It soon became obvious that the two types of nebulae were very different objects. The distant nebulae then became known as galaxies, whilst the term 'nebula' was retained for the interstellar dust clouds. These 'dust clouds' are the places where stars are created.

A Star Is Born

If an interstellar dust cloud is too warm, the speed of the atoms is too great for stars to form; but at temperatures around 10° kelvin the atoms can clump together under gravity. The clumps grow larger and can eventually combine and contract under gravity to form a protostar. A protostar cannot claim the status of a star until it begins to shine. A protostar consists mainly of the molecules in the nebula, namely, the lighter elements hydrogen (74 per cent) and helium (24 per cent), with the remaining 1 or 2 per cent comprising the heavier elements. As the protostar shrinks under its own gravity it heats up, and at some critical point it reaches a temperature at which nuclear reactions can take place. The most important reaction is the burning of hydrogen to produce helium, and once the temperature is high enough to fire this reaction the star begins to shine. The fusion of hydrogen into helium generates heat and light in the form of photons ('packets' of energy), and the radiation pressure produced by the hydrogen fusion is sufficient to prevent the star from collapsing under its own gravity. The young star enters a stable state where its temperature continues to rise until it becomes a main sequence star.

Nuclear Physics Provides the Answer

It is a curious fact that although astronomy deals with the science of very large objects and with the universe as a whole, in the 20th century it was found necessary to study the smallest entities in order to understand what was happening in the cosmos. The processes taking place inside the stars – the creation of new elements by the building up of atomic nuclei from protons and neutrons – could not be explained by chemistry; it required instead an understanding of atomic and nuclear physics.

There exist inside the stars conditions of temperature and pressure that could not possibly be created on Earth. Nuclear reactions are taking place. To begin with these were events that the astronomers could not fully understand, and it was necessary for them to enlist the help of the nuclear physicists to explain what was happening. A very good example concerned the Sun, our star. In the 19th century British physicist Lord Kelvin (1824–1907) calculated that the Sun could not shine for more than 20,000 years because by that time all its fuel would be used up. As we have mentioned in an earlier chapter, he had naturally assumed that the Sun was powered by chemical reactions when making this calculation of the star's lifespan. Kelvin knew nothing of nuclear physics. He was working before Albert Einstein proposed his theory of relativity, and thus did not know that the equation $E = mc^2$ proved that large amounts of energy could be created from small amounts of matter.

An Atomic Model

By 1900, knowledge of atomic theory was fairly well developed. The elements were tidily placed in their correct positions in the periodic table. The chemists did not at this time understand fully the internal structure of the atom, but they nevertheless had a good idea of the size of the atoms and molecules. Then, in 1910, Ernest Rutherford (1871–1937), a New Zealander who at that time was working at the University of Manchester, undertook his now-famous experiment in which he fired alpha particles at sheets of gold foil only a few atoms thick. The alpha particles were charged particles – the nuclei of helium generated by radioactive decay. The great majority of the particles went straight through the gold foil, but a small number were deflected in the process and an even smaller number, about one in 20,000, bounced back again from the foil. According to some researchers this rare event was like firing a bullet at a sheet of paper and having it bounce back again.

Rutherford was able to conclude that the atoms of

gold consisted mostly of empty space, and that nearly all the mass of the atom was concentrated in a nucleus at the centre. The nucleus carried a positive charge, but it was surrounded by orbiting electrons that carried negative charges. The electrons were attracted to the nucleus by electrostatic force, but instead of falling into the nucleus they orbited round it rather like a minute planetary system. The alpha particles carried a positive charge and they were therefore repelled by the positive charges in the nuclei of the gold atoms. Usually this effect produced a small deflection, but in very rare cases an alpha particle collided almost head-on with a nucleus. Whenever this happened the particle could be detected as it bounced off the gold foil.

Niels Bohr (1885–1962), a Danish physicist who was one of Rutherford's team in Manchester, formulated an alternative atomic structure. He suggested that the electrons in the atom were not like the planets in the solar system, but were instead confined to discrete orbits corresponding to specific energy levels. The electrons could change from higher-level to lower-level orbits and lose energy in the form of a photon. Conversely, the atom could absorb a photon and move an electron to a higher energy orbit. The great benefit of the Bohr atomic model was that it could be used to explain the lines generated by hydrogen and other elements in the spectrum of the Sun and other stars. In the Bohr model, only photons of particular frequencies could be emitted or absorbed by a specific element, with the spectrum of each element producing characteristic spectrum patterns. Astronomers could already identify elements by their spectral lines, but the Bohr atom explained how these lines were related to the atomic structure.

Another result of the new atomic theory was that it showed how it was possible to change atoms of one element to atoms of another. This could be achieved by the crude method of bombarding an atomic nucleus with fast-moving particles in the hope that one of them would strike the nucleus and knock out one or more of the protons. By this means the atom could be changed into an atom of a different element. It seemed to offer a solution to the age-old problem of the philosopher's stone, whereby base metals could be transmuted into gold. One case of transmutation was already well known. It was the radium clock used by geologists to date the age of the rocks. Uranium ore spontaneously gave out radioactivity in the form of alpha particles. In the radioactive process an atom of uranium was changed to one of thorium.

Not a Precise Science

To begin with, scientists struggled to understand the nature of light at the atomic level. Some experiments indicated that light consisted of particles, whilst others suggested it existed in the form of a wave. Austrian-Irish Erwin Schrödinger (1887–1961) was one of several physicists who helped develop the theory of quantum mechanics. He formulated an equation, known as the wave equation, that attempted to describe mathematically the way in which electrons and atoms behave.

The German physicist Werner Heisenberg (1901–76) developed a different theory called matrix mechanics. Later it was shown that both wave mechanics and matrix mechanics are really different mathematical approaches to the same theory. Heisenberg went on to suggest what became known as the uncertainty principle. It helped demonstrate the fact that both waves and particles are components of electromagnetic radiation, and therefore particles may act like waves in certain conditions.

Physics at the atomic level was no longer an exact science, and it was this that caused Albert Einstein to make his famous remark that *'God did not play dice with the universe.'* But on this occasion the world's greatest living physicist was proved wrong. Quantum mechanics defied common sense. It seemed totally illogical, but in the years between the two world wars the theory advanced to make many valid predictions about the world of the atom.

Energy from the Atom

It was soon established that there were theoretically two ways to produce energy from the atom. The first way was through nuclear fission. This is the energy that powered the atomic bomb, and it works by creating a very crude and uncontrolled explosion. Particles are released from radioactive uranium to strike other atoms of uranium, producing a chain reaction that results in a nuclear explosion. Nuclear fission is also harnessed in a more controlled fashion in a nuclear power station, whereby the heat emitted is used to produce steam to drive a turbine, thus generating electricity. The second way that atomic energy can be produced is through nuclear fusion. This is a process whereby nuclei are fused together. At the temperature of the Sun's core hydrogen nuclei are continually fused together to form nuclei of helium, and in the process energy is released in the form of heat and light. Despite extensive research, controlled nuclear fusion has never been achieved on Earth; we simply do not possess the technology. Nuclear fusion is used to create hydrogen bombs, but the reaction is an uncontrolled one. If the problem of controlling nuclear fusion could be solved it would give us a clean method of generating large amounts of energy for conversion into electricity.

The End of the Sun

The British astronomer Sir Arthur Eddington (1882–1944) was the first to suggest that the source of energy in the Sun was nuclear fusion. He asserted that the temperatures in the core were so high that hydrogen nuclei were stripped of their electrons, leaving a single proton. Four such protons were capable of fusing together to form a nucleus of helium (^{4}He) by changing two of the protons into neutrons. Other particles, called positrons and neutrinos, were given off during the transition. The mass of the helium atom created by this process is less than the mass of the four hydrogen atoms. The missing mass is converted into energy as given by Einstein's law $E = mc^2$ and it manifests itself to us on Earth as visible light and heat carried by the photons. Inside the Sun 4 million tonnes (3.9 million tons) of hydrogen are converted into helium every second, but this is no great cause for immediate concern – there is sufficient hydrogen in the Sun to last for another five billion years. After that time, however, when the hydrogen is finally all used up, the Sun will be left with a core of helium, and other thermonuclear reactions will begin.

In about five billion years time, when the hydrogen fuel is exhausted, the Sun will slowly expand until it becomes what we call a red giant. There are plenty of red giants visible in the sky, of which Betelgeuse in the constellation of Orion is the best known and most widely studied. When the Sun reaches this stage it will signal the final death throes of the Earth. The Sun will expand until it is so large that it will completely swallow up the orbits of both Mercury and Venus. The temperature at the centre of a red giant is high enough for the helium to burn. Once hydrogen burning ceases, the gravity will compress the core of the star, raising its temperature and density until they are high enough to start fusion of helium to form carbon and oxygen, thus providing a new source of energy to resist the inward pull of gravity.

It proved extremely difficult to work out how the carbon nuclei could be formed, and the astronomer Fred Hoyle (1915–2001) deserves much of the credit for providing the details. It requires three helium nuclei to form a nucleus of carbon, and in the process a vast amount of radiation is released as gamma rays. The triple collision between three helium nuclei is a far less likely event than the very common collisions between only two particles, but Hoyle proved that when the density and conditions were right then the carbon nuclei could be created in this way.

In its life as a red giant star the temperature and pressure in the Sun will be high enough for it to produce atoms of oxygen. There will be a small helium core at the centre of the Sun, about twice the size of the Earth, where carbon and oxygen are produced for a period of about two billion years. The Sun will remain as a red giant, slowly losing matter from its outer layers into the surrounding space and turning into what is called a planetary nebula, shrinking in size until it becomes a white dwarf.

Our Sun will end its life with all its remaining matter compressed into a sphere about the size of the Earth but with a mass of about 70 per cent of the original solar mass. The white dwarf is therefore an incredibly dense star; one teaspoonful of it would weigh about 5 tonnes (4.9 tons). But a white dwarf is a very dim star and for this reason comparatively few have been discovered, although one of our near neighbours has already reached this stage of its evolution. It has long been known that Sirius, the brightest star in the sky, is in fact a binary star, and the dark companion of Sirius is the nearest white dwarf star to the Sun. Thus the burning of the elements inside the stars is the process by which the elements higher up the periodic table are formed. In the case of the Sun, however, no element heavier than oxygen is created. Yet heavy elements are abundant in the Earth's crust and elsewhere in the universe. There must therefore exist another mechanism whereby they are created.

The Death of Larger Stars

Stars with much higher masses than the Sun follow a similar course of evolution until they reach the final phases of their life, but the more massive the star the shorter is its time on the main sequence of the H–R diagram (a system of star classification based on plotting the star's magnitude at a standard distance from Earth, against the star's colour due to its surface temperature). The period of time during which the star can create the lighter elements varies according to the mass of the star. The Sun will take about ten billion years to burn up all its hydrogen, but a more massive star, say 25 times the mass of the Sun, could exhaust all of its hydrogen in only about six million years. The larger star would take about half a million years to burn the helium to produce carbon. The carbon burns in a mere 600 years creating neon and oxygen. The neon burns in about a year and the oxygen in about six months. The silicon burns in a single day. Then there follows a very spectacular explosion as the star collapses on itself. The radiation pressure of the photons created inside the stars is the only thing that prevents them from collapsing under their own gravity. When the burning process ceases there is no pressure left to hold up the outer mantle of the star.

The star will cool rapidly by astronomical standards as the radiation pressure falls and the force of gravity takes over. In the 1930s the Indian-born American astronomer Subrahmanyan Chandrasekhar (1910–95) was able to show that if a star measured more than about 1.4 solar masses then it would have a very different future from that of the Sun. The evolution of the more massive stars is quite different from the less massive ones, and the full story took much longer to uncover. The first phase of the more massive stars is similar to that of the less massive stars. When they run out of hydrogen they are able to burn helium, and they expand to become red supergiants so large that they have a diameter about the same as the orbit of the planet Jupiter about the Sun. The gravitational field inside the larger star is immense, and the radiation pressure is not capable of supporting it. The star collapses under gravity and this drives the temperature up to more than a billion degrees kelvin. The burning of helium leaves behind carbon and oxygen, but at the incredibly high temperatures the nuclei of these elements are travelling at very great velocities – at a sizeable fraction of the speed of light. The star in turn creates nuclei of neon, silicon, phosphorus and magnesium. The time taken for all this to happen varies greatly according to the size of the star; some stages take far longer than others. The burning silicon produces iron. This is the most stable of the elements and no matter how high the temperature rises the iron will not 'burn'. The nuclei of iron cannot be changed to heavier elements.

Supernovae – the Cradles of Life?

The star begins to collapse into the core. The density of the core will be about 10^{17} kilograms per cubic kilometre, a figure that equates to a hundred times the mass of the Earth in every cubic kilometre. This is the density of the nucleus of the atom, for in fact the star has evolved to become what we call a neutron star. As the star collapses, all the matter falls back into the hot and dense body of the star creating an inferno brighter than anything else in the sky. The result is called a supernova. Such events, once rarely seen, are now observed almost routinely in other galaxies. Supernovae were used to discover dark energy in the 1990s.

The supernovae are an important class of star, but the full realization of their value is another story that was not revealed until the late 1950s and it will be told in the next chapter. It is sufficient for the moment to say that without the supernovae life on Earth or anywhere else in the universe could not exist.

Supernova Events

In the past one thousand years there have been only four sightings of exploding stars, or supernovae, in our own galaxy. The first was in 1006, and is not well documented. It was followed by another new sighting in 1054, when Chinese astronomers recorded a new star appearing in the Crab Nebula. The third was the well-documented star seen in the constellation of Cassiopeia by Tycho Brahe in 1572; it was so brilliant that it was visible in broad daylight for several weeks. The fourth supernova was seen by Kepler in 1604 and is again well documented.

Astronomers have waited patiently – and impatiently – for another supernova to appear so that it could be studied by modern methods and equipment. So far, they have had no opportunity to observe another supernova in our own galaxy. In 1987 they witnessed something almost as good – it was a supernova in our neighbouring galaxy, the Large Magellanic Cloud. It was so bright it could still be seen from Earth with the naked eye.

17

BEYOND THE
VISIBLE SPECTRUM

With the development of the radio telescope a new and vitally important astronomical tool became available. Now objects from deep in space, hitherto unknown, could be detected, unlocking more doors and answering more questions about the universe. At the same time the discoveries made by radio astronomy posed new challenges to our understanding of the stars and planets.

Karl Jansky (1905–50) was a physics graduate who joined Bell Laboratories in New Jersey, USA, in 1928. The company was developing the use of short radio waves for a transatlantic telephone service and was discovering that spurious radio signals, otherwise known as static interference, sometimes interfered with the transmissions. Jansky's job was to track down the source of the radio signals so that they could be eliminated. In 1931, using a rotating antenna that he had built, Jansky eventually found the source of the radiation. It seemed to originate from somewhere in the constellation of Sagittarius, in the centre of the Milky Way Galaxy. Exciting and intriguing though Jansky's discovery was, his employers refused his request to build a telescope to investigate the source of the radio waves further, since they were deemed not to be a significant problem for their planned communication system after all. Instead, Jansky was assigned to work elsewhere in the company.

It was left to another American, Grote Reber (1911–2002), a part-time astronomer living in Illinois, to explore the source of the radio waves originally located by Jansky. Inspired by Jansky's work, Reber built a primitive radio telescope in his back yard – the first one ever made. In 1936 he detected radio emissions from the Milky Way, confirming Jansky's earlier findings. Reber then went on to undertake a systematic survey of radio waves from the sky, and laid the groundwork for a major field of astronomical research.

Bigger Radio Telescopes

The early radio maps of the sky seem very crude compared with the optical maps of the time, because the far longer wavelength of radio means that the resolution is orders of magnitude lower. In the 1930s it was difficult to give accurate positions of objects detected in the radio spectrum, and it was impossible to link the emissions with specific stars. Much more sophisticated instruments were needed. World War II (1939–45) held back the progress of radio astronomy, but the use of radar indirectly helped with the development of the science.

After the war radio astronomy developed quickly. The Jodrell Bank Telescope in Cheshire, England, was first conceived in 1951. It was designed in the form of a large metal bowl with a diameter of 76 metres (249 ft), and it was the first radio telescope of such dimensions that could be fully rotated. The telescope bowl was originally designed in the form of a wire mesh, but by the time construction began the 21 cm (8.3 in) line (named after the wavelength of the radiation) had been identified in the spectrum of hydrogen and it was realized that this would become an important wavelength in radio astronomy. The design of the bowl was therefore changed from a wire mesh to a solid reflecting surface so that the 21 cm (8.3 in) radiation could be detected. The Jodrell Bank Telescope was functioning by 1957, in time to track the signals from the Soviet satellite *Sputnik I*, and it played a major part in the mapping of the radio sky.

'Little Green Men' and Pulsars

In 1967 a young astrophysics research student called Jocelyn Bell (b. 1943) was working at Cambridge University where she was studying for a PhD. The university had just finished building a primitive radio telescope consisting of 2 hectares (4.5 acres) of chicken wire. Instead of rotating like the Jodrell Bank Telescope, it remained fixed and it used the Earth's rotation to scan the skies. When the recordings of the signals received by the telescope were examined, something extraordinary was discovered. Jocelyn Bell and her colleagues noticed a regular pattern of pulses occurring – indeed, the pattern was so regular it seemed that the signal must be human-generated. The interval between the pulses was measured very accurately at 1.3373011 seconds. There was much speculation as to the cause of the pulses. The most likely origin appeared to be a nearby earthbound source, perhaps a machine from one of the locally situated electronics companies or possibly some new kind of radio-controlled agricultural equipment. The idea that the pulses originated somewhere on the Earth was soon proved to be unlikely, however, when it was discovered that the source of the pulses followed a sidereal cycle rather than a diurnal cycle. In other words, they were in phase with the rising and setting of the stars rather than with the Sun. The second most popular theory for the source was rather more imagina- tive than the first, and it was given the name 'little green men' because it implied that the pulses were being gener- ated from an extraterrestrial life form. When the popular press got news of the find they descended en masse to Cambridge to report on the story, hoping to print news of the first contact with an extraterrestrial civilization. The journalists received disappointingly cautious answers to their questions, however, and the little green men theory was soon abandoned.

But what did emerge eventually from the discovery was nearly as exciting for the astronomers. Attempts were made by the researchers to locate the position of the source in the sky as precisely as possible. Optical and other radio telescopes came to their aid and the source was soon located in the constellation of Taurus. Closer inspection showed it to be situated in the familiar cloud of gas called the Crab Nebula. Then, as the astronomers homed in even closer, the radiation was found to be coming from the very location of the supernova of the year 1054, the new star observed by Chinese astronomers more than nine centuries ago!

The fact that the radio pulses were associated with a supernova was an exciting find and it seemed very significant that the radio emissions should be generated not by what was seen as a 'new' star but by one in the final stages of its evolution. Because of its regular radio pulses, the word 'pulsar' was given to this new discovery. At that time there was no obvious mechanism linking regular radio pulses with a supernova. It was necessary to find an explanation.

Neutron Stars

As early as 1934 it had been suggested that there existed in space an amazing object called a neutron star. This was the remnant of a supernova created when a massive star became so compressed that it collapsed under its own gravity. The gravitational pressure at the centre of the star became so strong that all the neutrons became fused together to create a core of nuclear fluid of incredible density and with properties that could not possibly be simulated in a terrestrial laboratory. The neutron star was an unknown object but photons and neutrons were well understood and many of the properties of the neutron star, in particular its density, could be calculated from the theory of atoms and gravitation. The more the pulsar in the Crab Nebula was studied the clearer it became that the source did indeed have the properties expected from a neutron star. The case was examined and discussed at length by the astronomical world, and the only logical conclusion seemed to be that the object was indeed a rotating neutron star. For many months afterwards the supernova observed by the Chinese in 1054 became the most studied object in the night sky.

Just like the Earth, the neutron star had a magnetic axis that was inclined to its axis of rotation. As the star rotated, charged particles were accelerated by the magnetic fields to create a beam of radiation directed along the magnetic axis, and as this axis precessed the beam swept round like a searchlight describing the surface of a cone. The Earth happened to lie on the surface of that cone and consequently it received a pulse of radiation once on every rotation, in other words, every 1.3373011 seconds.

The discovery of a neutron star validated many theories. It also opened up two amazing possibilities. Both of these had been suggested some time before 1967 but both needed a neutron star to prove their case. One of these was the existence of conditions necessary for the formation of the heavy elements. The other was the existence of black holes. We will deal with heavy elements formation first, starting with an idea that originated about a decade earlier than the discovery of the pulsar in the Crab Nebula.

Solving an Elemental Conundrum

The astronomer Fred Hoyle knew that he was losing his case for the steady-state model of the universe, but in 1957 he co-published a paper, now known as the B^2FH paper, which he hoped would strengthen the argument for his theory. In the debate about steady-state versus Big Bang Hoyle had pointed out a weakness in the Big Bang theory. The theory stated that in the first three minutes of the primordial fireball it was possible to synthesize the nuclei of hydrogen and helium – these acquired electrons and became atoms, and they provided the basic material needed to make the stars. The Big Bang could also have synthesized more of the lighter elements in the periodic table, but what it could not have done was create the heavier elements. It was known that the Sun was powered by the fusion of hydrogen atoms to produce helium. But the Sun also contained elements much heavier than helium. The Earth and the other planets were also built from heavy elements – a fact that had been known for centuries – and they could be refined from rocks and ores. It was also well known that the magnetic field of the Earth was due to its molten-iron core.

The question was, where did the heavier elements come from if they had not been created by the Sun? The B^2FH paper described how the nuclei of the elements could be formed in the evolution of the stars. Red supergiants – in other words, very massive stars – can undertake nuclear syntheses beyond those of the Sun and they are able to create elements as heavy as carbon and oxygen. Hoyle recognized that there was a problem explaining the creation of the carbon atoms, but he was able to overcome this difficulty. A carbon nucleus contains six protons and six neutrons, so it can therefore be created from three helium nuclei. The problem is that although the collision of two particles is very common it is quite a rare coincidence for three particles to all appear at the same point at the same time. However, Hoyle was able to show that under suitable stellar conditions the triple collision is a common enough phenomenon, and also that sufficient collisions occur to create the carbon nuclei in the quantities required by observation.

Once over the 'carbon hurdle' then nitrogen, oxygen and the higher elements could be created with relative ease, but to create the heavier elements needed much higher temperatures and denser radiation. Hoyle reasoned that one place where these extreme conditions could be found was in the centre of a supernova. As the star collapses into a neutron star temperatures become so high that the energy exists for heavier nuclei to collide, and the nuclei of elements even higher in the periodic table are created. It is a sobering thought that supernovae like the ones we observe just a few times in every millennium in our own galaxy are the source of all the atoms of the heavy elements on our planet.

It is true that atoms can be split by collisions with other particles, but in general the atom is indestructible and it can exist forever. The human body is built up from complex organic molecules, but all those molecules are themselves composed of atoms. All the heavy atoms in our body have therefore been created inside a massive star or even in some cases a supernova. The atoms in our bodies are billions of years old – minute building blocks recycled through countless generations of living organisms. Through all this time the atoms remain 'as good as new'. Nor have they aged since the time they left the distant and long-forgotten supernova from where they were originally created. We all carry around inside us remnants of supernovae explosions that took place aeons ago and millions of light years away.

Elements across the Universe

Iron and the other heavy elements appear in great quantities both on Earth and in other planets of the solar system. A supernova event, like the ones we observe from time to time, seems a very slow way to form these elements, but when the years are counted in billions then it becomes clear that there have been a great many supernovae explosions throughout the history of the universe. We also need to ask some questions about how these elements have managed to cross the vast distances of space to reach our own planet and everywhere else in the galaxy. We know that it takes years to span the distance from star to star even at the speed of light, so how can all the elements cross the vast spaces between the stars to become relatively abundant around stars where planets are forming? The answer again is to be found in the vast aeons of time that have elapsed since the birth of the universe. If the atoms travel at only 1 per cent of the speed of light then it takes centuries rather than years to cross the spaces between the stars. But since the birth of the universe millions of centuries have been available for this to happen, and the atoms that formed the Earth could easily have travelled the vast distances across the galaxy – especially since the huge power of supernovae blasts would have contributed to the distribution of the elements over a wide region of space.

The whole sequence from the birth of the stars to the creation of the planets has gone through at least two long stellar cycles. The first cycle created stars of high mass that followed the normal evolutionary pattern to die eventually as supernovae. At their death, they manufactured atoms of the heavy elements. In the second cycle of stellar evolution planets like the Earth were able to build up a solid core out of the heavy elements made available by the dying stars, and were thus provided with all the elements needed for the evolution of life.

Black Holes

The pulsar in the Crab Nebula helped to prove the theory of the formation of the heavy elements. There was another mystery on which it also threw some new light. After Newton, but long before Einstein, an obvious consequence of gravitation had been pointed out. The more massive a body became the stronger was its gravitational field. It was possible, therefore, that if a star was large enough then radiation pressure would no longer be strong enough to prevent the total gravitational collapse of the star. The white dwarf star was very dense, but the neutron star was far denser. But there was an object even denser than a neutron star, an object with some astonishing properties. It was a collapsed star so dense that not even light could escape from its gravitational clutches. This dark object was given the name of a black hole. But could such a thing as a black hole exist? For many years astronomers thought such an object was physically impossible, but after the discovery of a neutron star they were forced to conclude that black holes were a possibility, and they tried to invent methods by which an object that emitted no light could be detected in the sky.

No Escape

There is another way to understand the phenomenon of a black hole. If a satellite orbits the Earth at a certain speed then its velocity is sufficient to keep it in orbit for an indefinite period. However, if the satellite is accelerated sufficiently it reaches what is known as its escape velocity, which means that it is moving fast enough to escape the Earth's gravitational pull completely. In the case of the Earth the escape velocity is about 7 miles per second (11.2 km/sec), but for a satellite at the Earth's distance from the Sun it is 26.1 miles per second (42 km/sec). The Earth already has a speed of about 18.6 miles per second (30 km/sec) around the Sun, so the additional speed to escape from the Earth's orbit and then from the solar system is about 7.5 miles per second (12 km/sec).

Every star also has an escape velocity from a given distance, and the greater the mass of the star the higher the velocity. With an extremely massive star it is necessary to take into account the effects of relativity. But Einstein pointed out long ago that the speed of light is the highest velocity that anything can ever achieve. The question asked by the astronomers was: is it possible to have a star so massive that the escape velocity becomes equal to that of light? If so, such a star would have no appearance to us because even light would be unable to escape from its clutches. The black hole was a very appropriate name – it was black because no light could escape from it and it was a hole because anything or anybody unfortunate enough to go too near it would be drawn in with no means of escape.

Evidence for the Black Hole

The discovery of a neutron star by astronomers made it seem far more likely that black holes existed in the universe. It may be impossible for us to see a black hole because it does not emit any radiation, but it is nevertheless possible to see the effects exerted on light and other objects by its massive gravitational field. It was not difficult for the mathematicians to work out the properties of the black hole. The space around it would be distorted by its very strong gravitational field. The bending of light predicted by Albert Einstein, for example, would be easily detectable. But when the properties of the space around the black hole were worked out it left an event horizon which is the limit of the black hole, but also an edge of the universe. The escape velocity at the event horizon equals the speed of light. To enter inside the event horizon of the black hole is to leave our universe and never to return. The distance of the event horizon can be calculated for a stationary black hole, and it is called the Schwarzschild radius.

Despite its mysterious nature, in astronomical terms, the black hole is quite a simple object. It has a measurable mass, which is usually large in terms of star masses. (There is insufficient gravity to compress all the mass if the star is less than about two or three times the mass of our Sun.) The black hole has spin, more precisely stated as angular momentum, and this means it has an axis about which it rotates and which causes it to be slightly flattened, rather like the Earth at the poles. It can also have an electrical charge.

At the time the first pulsar was discovered there were

still a few astronomers who doubted the existence of black holes. But there is now plenty of evidence to show that they do exist. Our Sun is a single star, but we now know that nearly half the stars in the sky are not alone. They have one or more companions, with the binary star being the most common arrangement. This means that there are many instances where a bright star orbits a dark star, and there are also some instances of where the dark star appears to be a black hole.

A good example of this arrangement is the X-ray source called Cygnus X-1. The visible star in the pair is a type B0 supergiant at about 8,000 light years from the Earth. It has a dark star orbiting very close to it. The star is a very strong source of X-ray radiation and the most likely explanation is that the radiation is caused by the presence of the black hole. The powerful gravity captures matter from the red supergiant and the matter orbits the black hole and it spins into a disc. Then the black hole draws the matter into itself, but in doing so the matter rises to a very high temperature and it becomes a very strong emitter of UV (ultraviolet) radiation. Some of the energy released by matter as it falls onto the black hole is converted to X-rays. The X-rays from Cygnus X-1 therefore imply that it is located very near to a black hole.

SETI – The Search for Extraterrestrial Intelligence

SETI, the Search for Extraterrestrial Intelligence, is an exploratory science that seeks evidence of life in the universe by looking for some evidence of its technology.

Radio waves can penetrate many parts of the galaxy where optical light cannot pass. There is a region of the radio spectrum called the 'water hole' that seems to be a logical frequency at which to search for messages from extraterrestrial intelligence, since signals at this narrow bandwidth are not known to occur naturally.

The search has been set up as an international SETI project. Its Project Phoenix was the world's most sensitive and comprehensive search for extraterrestrial intelligence. It began observations in February 1995 using the Parkes Radio Telescope in New South Wales, Australia. In August 1998 Project Phoenix moved to the upgraded radio telescope at Arecibo, Puerto Rico. Unlike many previous searches, Phoenix didn't scan the whole sky, but scrutinized the vicinities of nearby, Sun-like stars. Such stars were most likely to host long-lived planets capable of supporting life. Project Phoenix observed about 800 stars, all within 200 light-years' distance.

In a joint project with UC Berkeley, SETI is building the Allen Telescope Array in California, USA. This will be a SETI-dedicated array of 360 telescopes that will equal a 100-metre (328 ft) radio telescope. The first 42 antennas became operational in October 2007. In another SETI project, a large number of personal Internet-connected computers are used to process data. The SETI@home program is downloaded and runs when the computer is not in use, in the place of a screensaver. It downloads, analyses and then uploads the data.

There have been several false dawns, and as yet no evidence of civilizations other than our own has been found despite all our efforts.

18

BLACK HOLES, QUASARS AND THE UNIVERSE

Among the many mysterious objects in space, two in particular have fascinated scientists and non-scientists alike – even their names suggest something out of science fiction – and they may tell us more about the birth of the universe. Black holes are entities with such strong gravity that even light cannot escape from them. Quasars are bodies of indescribable energy that may exist at the very edge of the universe.

If we could build a spaceship to take us close to a black hole we would no doubt learn much more about these objects than we know at present. Of course, this is not a scenario within our current means of technology, and it is hard to envisage how it ever could be. Nevertheless, by studying black holes with the tools available to us we can still deduce much about these mysterious objects, even if we cannot get physically close to one of them. So, with such knowledge as we already possess, let us go on an imaginary journey to a black hole instead. We shall be visiting a single black hole – in other words, one with no orbiting companions. As we approach the black hole there will be changes in the space and time around it, although we may not be able to detect them due to relativity. So long as our spaceship keeps at a safe distance there is no immediate danger to us. We can orbit around the black hole just as we can orbit around the Earth.

A One-way Journey

Imagine now that we despatch an astronaut in a space pod to make a closer approach to the black hole and to try to get inside it. To help us track his movements the astronaut sends us a pulse every second. As he approaches the event horizon, the boundary of the black hole where escape velocity is equivalent to light speed, we will notice that the interval between the pulses becomes longer and longer. This is because the space around the black hole is distorted so that time passes at a slower rate. We observe that the space pod and the astronaut are subjected to what are called tidal forces by the gravity of the black hole. This means that the part of the space pod nearer to the centre of the black hole is subjected to a greater force than the part that is further away. The astronaut reaches a point where these forces impose a great stress on his pod. However, it is still not too late for him to return to the main spacecraft, provided he is still outside the event horizon, and can fire his rockets to travel at almost the speed of light. If he continues on his journey, then he and his space pod will be drawn inexorably into the black hole.

According to theoreticians the astronaut in the space pod will perceive events quite differently from those outside, again due to relativity. He is in his own space–time zone, distorted by the strong gravitational field which makes time pass relatively more slowly. Once the pod actually enters the black hole there is no way for the astronaut to communicate his findings to those outside and there is also no way for him to get out again into the universe he has left behind.

English mathematical physicist Roger Penrose (b. 1931) and English theoretical physicist Stephen Hawking (b. 1942) have shown that there is a singularity (a point at which density and temperature go to infinity) at the centre of the black hole, but we know nothing else about the space inside it. Space and time are stretched at the boundary of the hole such that the space inside is infinite. If the black hole is very massive then, although the gravitational field is stronger, the tidal forces are weaker. We would like to think that by entering into the black hole we could enter a new universe, similar to our own with stars and galaxies and habitable planets. And as we shall see later, there are even some who believe our own universe itself could in fact be located within a black hole. Science fiction writers have suggested that the black hole could be a wormhole – a connection between two very distant parts of the universe that would provide an ideal way of crossing the great distances between the stars. However, this idea remains in the realms of science fiction and is not a view held by astronomers.

Distant 'Quasi-stellar Objects'

In the late 1950s astronomers at Cambridge in England were putting together a catalogue of all the radio sources in the sky. Sometimes it was possible to follow up the radio source with an optical sighting, but the two are not always compatible and the radio emitters are not normally bright optical stars. In 1960 the American astronomer Allan Sandage (b. 1926) used the 5-metre (200-inch) Palomar telescope to study the star known as 3C 48 in the Cambridge catalogue. The star had some unusual features. Its spectrum showed strange emission lines that could not be identified from comparison with those of known elements. In 1962 another radio star called 3C 273 was discovered. It was unusual in that a long luminous jet was clearly visible in its optical image. Astronomers were puzzled by the strange emission lines that it also showed. In 1963, however, the Dutch astronomer Maarten Schmidt (b. 1929), working at the California Institute of Technology, was able to demonstrate that the unidentified lines in the spectra were in fact the well-known hydrogen lines. They had simply not been recognized due to the very high degree of redshift they were exhibiting. The objects were given the name 'quasars', an abbreviation for 'quasi-stellar objects'. Furthermore, they did not originate in our own galaxy; judging from their redshift they were travelling away from us at an unprecedented speed and they were therefore an unbelievable distance away. In fact it is no exaggeration to say that they were the most distant objects ever observed in the sky. The quasar 3C 273 was estimated to be at a distance of two billion light years.

The most distant galaxies studied before the discovery of quasars were not, after all, the end of the universe. The quasars were located in galaxies much further away, and because we were able to detect them from such a great distance it followed that they must have a very powerful source of energy. Once the first quasars had been identified, astronomers began to look for more objects with very large redshifts. They found that there were thousands of quasars visible with the most powerful telescopes. Redshifts of over 90 per cent have been measured, compared with measurements of only 2 or 3 per cent for the galaxies. It indicates that these quasars are many times further away. In fact, the most distant quasars are estimated to be 13 billion light years away. This means that we are observing them close to the time of the Big Bang and the origins of the universe. When the quasars came under close scrutiny it was discovered that they were located in galaxies, showing that the most distant galaxies were much further away than was originally thought.

When quasars came to be studied more closely it was found that their brightness was not constant, but the variations from the norm were usually very short-lived. Sometimes the quasars could flare up to be as bright as 100 galaxies – another indication that they must have a very compact but powerful source of energy to be so highly visible at such a great distance. There is only one power supply that could satisfy such a voracious appetite for energy. The evidence suggests that a quasar must contain a very massive black hole at its centre. The quasars are objects so powerful that they consume stars. Millions of stars must have been swallowed up to create the massive black holes that power the quasars, and when they devour a new star the energy released is so great that we see the flare of its death throes from the Earth. With the quasars we seem to have reached the very limits of the universe, but we have been wrong about this so many times that nobody can be sure. There was a time when the edge of the universe was thought to be the edge of the Earth, then it was thought to be the furthest part of the solar system, then it became the Milky Way. In the early 20th century the limit was assumed to be the most distant galaxies, now we think the quasars are at the end of the universe – but we cannot even say this with too much confidence.

Dusty Quasar Winds

Using the Spitzer Space Telescope infrared spectrograph instrument, scientists found a wealth of dust grains in a quasar called PG 2112+059 located at the centre of a galaxy eight billion light years away. The grains, which include sapphires, rubies, peridot and periclase (naturally occurring in marble), are not normally found in galaxies without quasars, suggesting they might have been freshly formed in the quasar's winds.

These findings are another clue in an ongoing cosmic mystery: where did all the dust in our young universe come from? Dust is crucial for efficient star formation as it allows the giant clouds where stars are born to cool quickly and collapse into new stars. Once a star has formed, dust is also needed to produce planets. Theorists had predicted that winds from quasars growing in the centres of distant galaxies might be a source of this dust. While the environment close to a quasar is too hot for large molecules like dust grains to survive, dust has been found in the cooler, outer regions. Astronomers now have evidence that dust is created in these outer winds.

More than Three Dimensions

The universe is often described as a four-dimensional space–time continuum, consisting of the well-known three spatial dimensions that we experience (length, width and depth), plus another dimension – the time dimension – that has a special significance in Einstein's special relativity theory. Ignoring the time dimension for the moment, there is some evidence that the three spatial dimensions of the universe do not extend to infinity; the universe may be confined in a closed space in the same way that the sphere of the Earth is enclosed by its surface.

Imagine a two-dimensional being at the North Pole. We have drawn a set of concentric circles – lines of latitude – for him or her about the pole. These circles get larger and larger as they progress south, eventually reaching a maximum width at the equator. From there, the circles grow smaller and smaller until they reach the South Pole, at exactly the opposite point from the equator as the North Pole. If our two-dimensional traveller heads south from the North Pole he will eventually reach the South Pole, and if he still keeps going in the same direction he will eventually come back again to the very point he started from at the North Pole.

Now imagine we stop the Earth's rotation and delete the latitude and longitude lines so that the North and South Poles do not hold any privileged position on the sphere. A voyage around the world can start from any chosen point on the surface and the traveller, provided he can keep on a straight course, can travel round the world and arrive back at the point from where he started. The traveller can claim that his starting and finishing point is the centre of the world. We, as three-dimensional beings, can see that the claim is not valid; any point on the surface of the sphere can claim to be at the centre. We can see the whole sphere and we know that although the traveller's world does have a centre, it is not on the surface of the Earth. The centre does not have a latitude or longitude. It is in a third dimension at the centre of a sphere. If our two-dimensional being is clever enough, he may be able to deduce that his space is curved. The traveller defines a straight line as the shortest distance between two points; it is what we know as a great circle. He knows that in a flat space the angles of a triangle add up to 180 degrees or two right angles. He draws a very large triangle using great circles for the sides and discovers that the angles add up to more than 180 degrees. If he takes two points on the equator, separated by 90 degrees of longitude, and joins them both to the pole, then he can create an equilateral triangle whose angles add up to three right angles. He would also find, if he were to draw a very large circle, that its circumference is less than the value given by the familiar formula $2\pi r$. The radius of the equator, for example, is the same as the radius of the Earth from our viewpoint, but for our two-dimensional being the radius would be a line from the North Pole to the equator and the circumference of the circle would not be multiplied by the radius, it would be four times the radius. In theory, the traveller could use this result to calculate the size of his spherical world.

Thus the surface of the Earth can be thought of as a two-dimensional world needing a third dimension to explain it. The universe, excluding the time dimension, is a three-dimensional world, but if it is confined within a boundary then it needs a fourth dimension to explain where the boundary lies. We can define a three-dimensional 'spherical' space as the boundary of a four-dimensional super-sphere. We are not capable of envisaging where the centre of this super-sphere lies, but mathematicians can deduce many of its properties. The circle of radius r, for example, can be represented by the equation $x^2 + y^2 = r^2$, the sphere with the same radius can be written as $x^2 + y^2 + z^2 = r^2$ and the four-dimensional 'super-sphere' of radius r can be expressed by the equation $x^2 + y^2 + z^2 + w^2 = r^2$

where the four axes, (x, y, z, w) are all at right angles to each other. From this the mathematicians can work out all the properties of the super-sphere. If we cut it with the plane '$w = 0$', for example, then we find the three-dimensional section is the sphere $x^2 + y^2 + z^2 = r^2$. We can work out all the properties of the super-sphere such as volumes, areas, distances and angles.

Now we are ready to send out two intrepid observers from our planet to the most distant quasars. We will remain safely on Earth. When our observers arrive at their destination they, and we, will measure the angles of the triangle we have formed. If the angles add up to more than two right angles then this shows that the universe is closed. If we know the distances then we could make a measure of the curvature of the universe and we could calculate how far our two distant observers would need to travel to get back again to the point from where they had started. If the angles of our triangle add up to exactly two right angles then the universe is perfectly flat and it extends to infinity in all directions. If the angles add up to more than two right angles then the universe is divergent; the two-dimensional analogy is the surface of a saddle where the perimeter of a circle, for example, is greater than 2π times its radius.

Where Are We?

There are many theories about the nature of the universe we live in, and one of the most fascinating is that the universe itself is contained inside a black hole. The Schwarzschild radius of a black hole is given by the simple formula $r = 2GM/c^2$. Here G is the gravitational constant, M is the mass of the black hole and c is the velocity of light. Thus it is the mass of the black hole, the only variable in this equation, that determines the radius and therefore the boundary of the black hole. A black hole with the mass of the Earth, for example, would be about the size of a cherry, and it hardly needs saying that it would be an incredibly dense object. A black hole with twice the mass of the Earth would have a radius twice that of a cherry; it would be eight times the volume, and the density of the matter in it would still be very high but only one quarter that of the smaller black hole. If the black hole had the same radius as the Earth it would be incredibly massive, but the density would be only 5×10^{18} times the density of the cherry-sized black hole. For very large black holes, with radii measured in millions of light years, the density of the matter inside comes down to manageable proportions. It is not necessary for a black hole to contain incredibly dense matter inside its horizon. The next question we need to ask therefore is: 'what would be the radius of a black hole with a mass equal to that of the whole universe?' If we knew the mass of the universe then we could calculate the Schwarzschild radius of such a hole very easily, and we know that the radius would be a distance of astronomical proportions. This would mean that the whole universe was contained inside a black hole. Nothing can escape from the universe. Anything trying to escape even at the speed of light would be drawn back to it by the gravity of all the mass in the universe. If we really were inside this black hole then how would it appear to us? It has to be admitted that it may well appear to be very similar to the universe we do actually observe.

The 'black hole' description of the universe complements the idea of a three-dimensional but finite universe quite well. We would not be able to find the edge of our universe. We could send off a light beam in what we think is a straight line to the edge of the universe. After travelling for a few billion light years we think that our light beam has travelled in a straight line, but the four-dimensional being looking from the outside can see that it has followed a curve. He or she can calculate how long it will take for the beam to travel in a great circle and to come back from the opposite direction to the point from where it started. Looking in the direction of the light beam we are unaware that it is not straight; light from any bright object from the sky behind is bent in exactly the same way by the gravitational field of the universe inside the black hole.

Similar arguments can be used to answer the question of where the universe was first created. Where should we point our telescopes to see the place where the Big Bang took place? The answer is that we can point the telescope anywhere in the universe and we are looking towards the place where it all began. It seems a very unsatisfactory answer, but from what we are able to deduce about the universe it does not seem to matter what point we choose, for to look at the universe from that point would show the same results as we see from our own vantage point. It is true that we see everything rushing away from us, faster and faster the further away it is. It can be argued from our three-dimensional minds that we are at the centre of the universe, where the Big Bang took place, otherwise we would see galaxies and quasars at the edge of the universe travelling at different speeds away from us with the direction of the centre of the Big Bang as the point where those speeds are the greatest. But this is not the case. We are like the raisins in a cake expanding as it is cooking. Each raisin 'sees' all the other raisins rushing away from it in every direction, but it does not have a privileged position in its space.

19

STEPHEN HAWKING

Exploring the Boundaries of Space

For many years our understanding of the universe was based on the interpretation of data painstakingly collected through countless hours of observing the heavens with optical and, later, radio telescopes. But then came a major advancement: ground-breaking astronomical discoveries about the universe brought about with the aid of complicated mathematics.

There was a public lecture in the university town of Cambridge. The speaker had to be lifted onto the stage in his wheelchair by several assistants. He was wired up to a computer and a voice synthesizer was plugged into a public address system. On the arm of the wheelchair were a set of computer discs containing text and diagrams for the lecture. Pictures and diagrams appeared on the screen. The speaker addressed his audience by tapping the controls on a keyboard to operate the speech synthesizer. A stilted, metallic, computerized voice rang out with a slight American accent. There were gaps between the words and sentences but the presentation was still fully comprehensible to the audience. Sometimes the speaker raised his head. Sometimes he lifted his eyebrows and turned to face the audience with a wicked grin. The lecture lasted for about 40 minutes and was greeted with rapturous applause. Professor Stephen Hawking acknowledged the ovation.

The Birth of a Special Astronomer

It used to be said that Isaac Newton was born in the year that Galileo died. This claim can only be made because the English and the Italians were using different calendars at the time; thus when Newton was born it was still 1642 on the Julian calendar in England but it was 1643 on the Gregorian calendar in Rome. Stephen Hawking was born on 8 January 1942 – 299 years after Isaac Newton was born and 300 years to the day after Galileo died. His career parallels that of both his predecessors.

Hawking's birthplace was at the 'other place'; the university town of Oxford. This came about because his family moved out of Highgate in London at the outbreak of World War II. The family moved back to Highgate when the war ended and in 1950 they moved on again to live in St Albans. It was there, at the local grammar school, that Stephen Hawking received his secondary education and where he achieved the academic standard necessary to return to his birthplace at Oxford as an undergraduate student. As a bright student he was able to enjoy the social side of undergraduate life to the full and still manage to pass his examinations. By the time he approached his finals he had decided that he wanted to devote his life to an academic career. He was well aware of the rapid progress being made in astronomy and cosmology and he wanted these subjects to be the mainstays of his career. There is a story about his viva interview at Oxford, to determine his class of degree. *'If you award me a first,'* he said, *'I will go to Cambridge. If I receive a second, I shall stay in Oxford, so I expect you to give me a first.'* Did Oxford want to get rid of Stephen Hawking or were they perfectly fair in their dealings? It is hard to believe that the examination board took his words to be any more than a joke but the outcome was that they awarded him first class honours.

In the astronomical community the man that Hawking admired above all others was the blunt, outspoken and brilliant Fred Hoyle (1915–2001) at Cambridge. Hoyle was the main reason why Hawking chose Cambridge over Oxford for his postgraduate career when he had such strong connections with the latter university. The other reason for choosing Cambridge was that he had already decided that he wanted to follow a career in cosmology and at that time Cambridge was much better placed to offer him that career. He quickly settled into postgraduate life, but it was not all roses. When he began his PhD he quickly discovered that his knowledge of mathematics was inadequate. To follow the research he had chosen to undertake he had to work hard to master the tensor calculus – a necessary requirement to be able to understand the work of Einstein and the complex calculations required for general relativity.

A Bitter Blow

It was at Cambridge that Stephen Hawking met a young lady called Jane Wilde. She found him an eccentric and fascinating character, but she also soon discovered that there was to be a tragic side to him.

> *There was something lost,'* she said later. *'He knew something was happening to him of which he wasn't in control.*

Hawking was 21 at this time. He knew that he was suffering from a medical problem and so he arranged to have an examination and a professional diagnosis. The result of the examination was not good news. The diagnosis showed that he was exhibiting the early symptoms of a rare disease called amyotrophic lateral sclerosis (ALS), better known in Britain as motor neuron disease. It affects the nerves of the spinal cord as well as the part of the brain that controls the movement of the muscles and limbs. It was a terrible thing to happen to a young man in his prime, and Stephen Hawking knew that his problem could only get worse. There was no cure for the condition. According to the statistics of those suffering from the disease he had very few years left to live. But although he could never get better, there was a small consolation. The motor neuron disease did not affect his brain. He was still capable of doing his research and following the academic career he wanted.

Overcoming a Tragedy

In July 1965 Hawking married Jane Wilde. She knew that he was by this time suffering from crippling motor neuron disease and was fully aware of what she was taking on. Nevertheless she was firmly of the belief that he had a great future in spite of his growing disability. By this time Hawking had taken to using a walking stick to help himself to get around. Soon afterwards his speech deteriorated and also his mobility. He had to change his stick for a pair of crutches. For a few years the illness progressed slowly, but his speech continued to deteriorate and there came a point when only his closest associates could understand what he was saying. His mobility became seriously affected and from 1969 he was issued with a standard National Health Service three-wheeler invalid carriage to get around. However, by 1974 he found it more convenient to use an electric wheelchair. By 1985 his speech had deteriorated so much that, after struggling on for a decade, he was obliged to use a speech synthesizer to give his lectures and to communicate with other people. This was actually a great help to him and he quickly found that he could communicate faster and better with the speech synthesizer than he had been able to do for many years without it.

Hawking is an exceptional case as far as motor neuron disease is concerned. Not only has he lived far longer than is usual for someone suffering from this condition, but he is still solving new and complex problems in cosmology at an age when many mathematicians are seen as well past their prime. We cannot help but admire a man who has pursued a demanding fulltime career whilst having to compete with the ravages of motor neuron disease. There are many anecdotes of his reckless wheelchair driving, his sense of humour and his methods of coping with his disability, but he will no doubt not wish to be remembered for his illness but by his contribution to cosmology. In his chosen field of study Stephen Hawking became the most charismatic figure in the second half of the 20th century.

Forging a Different Path

As a young man Hawking had no difficulty using a telescope, but the direct observation of the universe was not his first priority. He was quite content simply to read about new observations in the astronomical press. He knew from the outset that he wanted to be a cosmologist. He was much more interested in aspects of astronomy other than direct observations. He was fascinated by the links with nuclear physics and by the mathematics of the remarkable objects such as neutron stars – and in particular black holes – that were very much at the cutting edge of cosmology in his time. He knew as much as most physicists about general relativity, and he also knew about the strange world of quantum mechanics. But whereas Einstein wanted nothing to do with quantum mechanics, Hawking was keen to bring relativity and quantum mechanics together, and this was an ambition in which he eventually succeeded.

He always entered enthusiastically into debate and was never backward in stating his own opinions. One of the best examples of his approach occurred at a meeting of the Royal Society in the early 1960s. At this time Hawking was just a young researcher, but at the meeting he challenged a statement made by his idol, Fred Hoyle. Hawking claimed that one of the quantities that Hoyle had specified in his equations as convergent was actually divergent. Hoyle asserted again that the function converged, but the younger man stood his ground. Hoyle was furious at being challenged in public but he later accepted that Hawking's proof was correct. They were both scientists. It was the truth that mattered. Hoyle held no grudge against Hawking, and any friction between the two cosmologists was quickly forgotten.

Fred Hoyle and the Steady-state Theory

Fred Hoyle (1915–2001), Hermann Bondi (1919–2005) and Thomas Gold (1920–2004) are the names chiefly associated with the steady-state theory of the universe, but in fact another British astronomer, James Jeans (1877–1946), put forward the idea in the 1920s. The steady-state theory, a competing cosmological view to the Big Bang, proposes that new matter is constantly created as the universe expands. After the discovery of background radiation thought to have come from the Big Bang, the steady-state theory quickly lost ground.

However, Hoyle went on to make a notable contribution to astronomy with his work on the evolution of the stars and in particular the way elements were created inside the stars. At first it was difficult to explain how the heavy elements could be created, but he showed that the necessary conditions for this to happen existed inside an exploding supernova. His classic paper on this subject, known as the B²FH paper, was published with William Fowler (1911–95), Margaret Burbidge (b. 1919) and Geoffrey Burbidge (b. 1925) in 1957.

Hawking and the Laws of Thermodynamics

Hawking was very intcrested in the discovery of quasars, and he was one of the scientists who believed that they must be powered by massive black holes in order for them to give out so much energy and to be visible at such a great distance. In the 19th century a branch of physics known as thermodynamics was developed to help improve the efficiency of the steam engine. Thermodynamics had fallen out of fashion by the time of Hawking but it still contained some interesting ideas, and one of them was the concept of entropy. It is difficult to give a precise definition for entropy, but the closest we can get is to call it the degree of disorder. Left to its own resources entropy always increases, and it requires the expenditure of energy to reduce the disorder. In the 19th century thermodynamics suggested that as the entropy increased the universe would slowly run down until it was the same temperature everywhere, then there would be no available energy left to reorganize it.

Through his studies Hawking knew that black holes could not emit any radiation and it followed that they could either remain the same size for ever or grow larger by swallowing up extra matter – but a black hole could not shrink in size. The rule was almost the same as that of the second law of thermodynamics wherein the entropy of a closed system could only remain the same or become greater. Hawking then realized that the mathematics of the two was identical, and he came to the conclusion that the surface area of the black hole was a measure of its entropy. Hawking also knew that the temperature of a black hole was in inverse proportion to its mass. He realized that the laws of thermodynamics suggested exactly the same thing. The smaller a black hole the greater its temperature. When he came to consider very small black holes – smaller in size than the proton – he realized that they must be very hot indeed.

Hawking next turned his mind to singularity theory. A singularity is a point at which physical quantities such as density and temperature become infinite, and there had long been a theory that for these quantities to reach infinity was just not possible. The saying that 'Nature abhors a singularity' was the opposite of the saying that 'Nature abhors a vacuum', but they were also parallel statements. Hawking worked closely with Roger Penrose (b. 1931) on the theory, and both men agreed that singularities must exist at the centre of a black hole and that they also existed in the first instant of the Big Bang. One argument they put forward was that it was impossible to see beyond the horizon of a black hole and therefore the existence of a singularity inside it did not matter because there was no way it could affect the rest of the universe.

Black Holes Meet Quantum Mechanics

Then Hawking's researches came up with an astonishing result. It had been known since the first theories of the black hole were expounded that nothing could possibly escape from inside the event horizon. Hawking, however, produced a theory to show that this was not always true; under certain conditions a black hole could in fact emit radiation. He arrived at this result after applying the laws of quantum mechanics to the problem. It was also known that under certain conditions particles could be generated from pure energy.

Astronomers have found evidence for what is called 'virtual particle' production. Nature allows pairs of 'virtual' particles to form spontaneously at any point in space. Each pair consists of a particle and its antiparticle, for example an electron and a positron, a proton and an antiproton, or a pair of photons (the photon is its own antiparticle). Normally, when the particles are formed, they annihilate each other after a time interval of about 10^{-21} seconds. If the particles appear near the event horizon of a black hole, however, then the gravitational field of the black hole changes the virtual particles into real particles. One of the particles is captured by the black hole and is never seen again. The other has sufficient energy to escape. Thus, seen from a distance, the black hole seems to emit a particle. It has also swallowed a particle so we might think that it has grown in size. In fact the reverse is true. The black hole has expended energy on the virtual particles and the net result is that it has lost energy (or mass if you prefer to think in terms of mass). Thus the spontaneous pair production process causes the black hole to slowly wither away.

Even stranger things could happen when black holes and quantum mechanics come together. Hawking tried to work out the properties of very small black holes. Readers will remember that a black hole with the same mass as the Earth would be about 1.7 centimetres (0.7 in) in diameter. In an earlier chapter we also saw how a very large black hole can have such a low density that the whole universe could in fact be a black hole. Now we must consider the extreme opposite: a black hole smaller than a proton. As the mass of a black hole is reduced so is the radius, but although the mass reduces the density increases as the radius decreases. Hawking calculated that a black hole weighing about a billion tonnes, the size of a mountain on Earth, would be contained inside a sphere smaller than a proton! The only conditions where the pressure was high enough to compress something the size of a mountain into a space the size of a proton was during the early phases of the Big Bang. He reasoned that there could be many black holes of this size left behind by the Big Bang when the universe was first created. These microscopically small black holes were amazing and mind-boggling objects. How could the mass of a mountain be compressed into a single proton? The black hole evaporates slowly but the nature of its final demise is in dispute. Every black hole has an infinite singularity at its centre; we do not know if this singularity creates a large explosion or if the black hole disappears peacefully. Quantum theory predicts that the black holes were not always stable and that they could explode spontaneously with the release of an incredible amount of energy. Hawking thought they were nevertheless sufficiently stable to still be around billions of years after the Big Bang.

Cosmologists Take Centre Stage

Interpreting the observations and images from space often requires the application of very exotic physics and esoteric mathematics. Hawking and Penrose wanted very much to find and observe an exploding black hole to prove their theories. They knew that by measuring the frequencies and amplitudes of the radiation and by a careful examination of the spectrum they would learn much about its history. They became convinced, much against current thinking at the time, that it was in fact possible to obtain information from a black hole about what had fallen into it. Thus in the late 20th century and into the third millennium the origins of the universe became the realm of the cosmologist more than any other kind of astronomer.

Hawking has published a great number of academic papers, but he has also written for the popular press. In 1979 he was the co-author with German physicist Werner Israel (b. 1931) of the book *General Relativity, an Einstein Centenary Survey*. Hawking's best-known publication, however, is *A Brief History of Time* published in 1988 and which remained for several years in the list of bestsellers. It was this book that helped solved the financial problems associated with Hawking's long-suffering family and the growing expense needed for his nursing and medical care. In 2001 his book *The Universe in a Nutshell* was published; it became another bestseller, providing glossy full-colour pictures of the latest ideas in cosmology. Stephen Hawking is the only man in modern times to approach Albert Einstein as a scientific icon, and it may be a long time before we see his successor.

Hawking is the Lucasian Professor of Mathematics at the University of Cambridge – the chair occupied by Isaac Newton in the 17th century. His ambition is very similar to that of Albert Einstein: he seeks to find the holy grail of science, a theory to unify all the existing theories into something that describes the whole of the universe.

20

ASTRONOMY IN THE SPACE AGE

By the late 1950s, a momentous new development in space exploration had been achieved: at that time it was not only possible to send spacecraft into orbit but also for humans to experience the great void of space at first hand. The space age had begun. It even became possible for people to set foot on another world. Soon scientists were looking beyond the Moon and sending probes to the planets in the remote reaches of the solar system.

The second half of the 20th century saw the first attempts by engineers and scientists to explore the solar system with the use of spacecraft. In 1957 the world was astounded when the Russians announced that their first satellite, *Sputnik 1*, was in orbit around the Earth. The Russians soon followed this achievement by becoming the first nation to send a living animal, the dog Laika, into orbit. On 12 April 1961, the Russian spacecraft *Vostok 1* orbited the Earth in a flight lasting 108 minutes with the cosmonaut Yuri Gagarin (1934–68) on board. Gagarin thus became the first human to enter space.

Landing on the Moon

As might be expected, the Americans were severely taken aback by these Russian space achievements, and in an attempt to claw back the initiative the US president John F. Kennedy (1917–63) announced a very ambitious project to put a man on the Moon by the end of the decade. There followed the so-called space race between America and Russia, with the early honours going to the Russians. In 1959 the Soviets achieved the first fly-past of the Moon, followed by the first hard landing on the Moon's surface, and then the first orbit of the Moon. During the orbit, the probe took remarkable photographs of the Moon's far side – it was the first time this view of the Moon had ever been seen because it always presents the same face to us as it orbits the Earth.

The financial cost of the space race was extremely high, and for this reason the Soviets could not hope to stay ahead of the Americans for very long. The Americans soon had satellites orbiting the Earth as well as piloted orbital flights. By the middle of the decade the American *Apollo* programme was well under way. On 20 July 1969 Neil Armstrong became the first person to step on the surface of the Moon. The *Apollo* programme achieved most of its aims, and out of the seven lunar missions (*Apollo 11* to *Apollo 17*) only the ill-fated *Apollo 13* did not reach the Moon. *Apollo 13* remains as the greatest drama in the early history of space flight, after the dramatic return of the astronauts to Earth following a failure in the command module.

When samples of Moon rock were returned by the *Apollo* missions and the close-up views of the lunar surface were studied there was a great scientific return for the money spent on the space effort, but there was little of direct commercial value. After the Moon landing the next step in the exploration of space no longer involved piloted missions. Life support systems were very costly and heavy, and it was far more efficient and less dangerous to explore the solar system by means of robot spacecraft and to transmit the findings back to Earth.

Exploring Mars

Mars was the next target. In the 1960s several of the *Mariner* missions mapped practically the whole of the surface of Mars and produced strong evidence that the surface had supported sustained water flow at some time in the past. They were followed in the mid-1970s by the *Viking* landers which touched down on Mars to become the first craft to send back views from the planet's surface.

As early as 1877 the Italian astronomer Giovanni Schiaparelli (1845–1910) had studied Mars through his 20-centimetre (8 in) telescope and discovered what he thought were sets of lines criss-crossing the surface. The American astronomer Percival Lowell (1855–1916) examined this suggestion further, and by the end of the century he had produced an image of the red planet showing a network of canals, built perhaps to carry water from the poles to the Martian desert for irrigation. This imaginative interpretation of the geological features of Mars developed by Schiaparelli and Lowell was soon discredited, but when the *Viking* orbiter crafts produced the first detailed maps of the Martian surface they found valleys that could only have been created by running water at some time in the remote past. There were also plains and craters on the surface, as well as mountains and extinct volcanoes. One volcano, Olympus Mons, is far bigger at 15 miles (25 km) in height than any others in the solar system. Mars also has canyons even greater than the Grand Canyon in the USA.

Martian Adventures

At the end of the current decade we will know far more about Mars than we know at present. Whilst Mars has been a primary target for planetary exploration, unfortun-ately many missions have proved unsuccessful, contact with the spacecraft being lost at launch, en route or crash-landing onto the surface.

However, orbiting crafts and landers are both providing more and more data to piece together the puzzle of Mars. The *Reconnaissance* orbiter was launched in August 2005 and is now providing more detailed mapping of the Martian surface from orbit.

The *Opportunity* and *Spirit* rovers have spent four years exploring Mars and their examination of surface rocks have provided the best evidence yet that Mars was once covered by oceans of liquid water. Each rover has travelled for several miles, and in 2006 *Opportunity* reached the edge of Victoria Crater, after spending many months exploring the smaller Endurance Crater. The rover had to shelter in a crevice whilst waiting for a large dust storm to clear. A safe path was found, and *Opportunity* entered into Victoria Crater. It is hoped that the crater will show evidence of how it was formed, possibly providing clues to the ancient surface history of Mars itself. The far rim of the crater, lying about 800 metres (2,625 ft) away and rising about 70 metres (230 ft) above the crater floor, can be seen in the distance. The alcove in front has been given the name Duck Bay.

The *Phoenix* lander is currently en route, and will search likely sites for signs of water. Two further *Scout* missions are planned to arrive in 2013 and 2018 and they will eventually provide much better exploration data. The geological survey of the whole of Mars will be complete within a few years, by which time the scientists will attempt much closer surveys of the more interesting areas. Geologists on Earth want a sample of Martian rock and the Mars Science Laboratory, to be launched in 2009, will help to provide one. It is hoped that a sample can be brought back to Earth by 2014.

Exploring Venus

Venus was the next planet to be explored by probes. The planet had traditionally been thought to be the one most similar to Earth, albeit with a slightly smaller mass and a much warmer climate. The Russian *Venera* probes reached the surface of Venus in the 1970s to early 1980s and were able to determine conditions underneath the thick layer of cloud that obscures the planet. The temperature was a searing 470 °C (880 °F), and the atmosphere was 96 per cent carbon dioxide and about 4 per cent hydrogen. It was also about 90 times denser than the Earth's atmosphere. None of the *Venera* probes lasted longer than about two hours, succumbing quickly to the extreme heat and pressure at the surface. There had been much volcanic activity on Venus, producing large quantities of sulphur in the atmosphere. Indeed, the dense clouds that enveloped the planet were found to consist mostly of sulphuric acid. Radar surveys showed many gently rolling hills on the surface, many volcanoes and two 'continents'. There were many craters including a huge impact crater on the surface, christened Klenova, which was 88 miles (142 km) across. It was a great disappointment to discover that the surface of the planet of love could hardly have been a more hostile environment.

Exploring Mercury

In 1974 the *Mariner 10* probe reached Mercury and went into an orbit around the planet that enabled three close approaches to be made. The probe mapped a remarkable 45 per cent of the surface during these orbits. The surface proved to be heavily cratered like the Moon. Unfortunately the planet was no more hospitable than Venus. The very thin atmosphere consisted of hydrogen, helium, sodium, potassium and a trace of oxygen. As a result of the thin atmosphere the daytime temperatures reached about 350 °C (662 °F), but at night they fell to about 170 below zero. Features examined included the Caloris Basin, about 808 miles (1300 km) in diameter and surrounded by a ring of mountains. It was also possible for the probe to measure the length of the day on Mercury. The year on Mercury has long been known to be the equivalent of 88 Earth days, but the solar day turned out to be 176

Earth days – two Mercurian years! The first fly-by of NASA's *Messenger* mission, launched in 2004, was achieved in January 2008, with two more fly-bys to come before it enters Mercury's orbit in 2011.

Exploring Jupiter

The development and launch of the very sophisticated *Voyager* probes in the 1970s meant that exploration of the outer planets of the solar system became possible. The two *Voyager* probes visited Jupiter in 1979 and Saturn in 1980 and 1981, with *Voyager 2* travelling to study both Uranus and Neptune during the late 1980s. These missions were followed by the *Galileo* spacecraft's visit to Jupiter, and *Cassini's* sojourn at Saturn in later decades. To save fuel, these probes were designed to reach the outer reaches of the solar system by using the gravitational pull of the other planets to produce a sort of slingshot effect, helping to propel them on their journey. Very sharp and interesting images of the gas giants and their satellites (moons) were transmitted back to Earth. Clearly visible on the surface of Jupiter – the biggest planet in the solar system – is the Great Red Spot. This feature is so large that it was seen in the 17th century. It is about 15,500 miles (25,000 km) long by 7, 456 miles (12,000 km) wide – large enough to swallow two Earths. It takes about six days to make a rotation, and although it changes shape over a period of time it has been a permanent feature on Jupiter for over 300 years. It is best described as a colossal hurricane or typhoon. The energy needed to maintain it comes from inside the planet, and it is part of an ever-changing atmosphere around Jupiter.

In 1993 there was great excitement when it was realized that the comet Shoemaker-Levy 9 was approaching Jupiter and that it would strike the planet in 1994. The comet was broken into more than 20 pieces by the gravitational field of Jupiter and the fragments struck the planet between the 16th and the 24th of July 1994. The Hubble Space Telescope and other large telescopes were brought to bear on the event and it was also observed and recorded by the world's astronomers. The impacts punched dark, giant asymmetric holes in Jupiter's atmosphere, which took several months to dissipate.

The *Galileo* probe was able to make a spectral analysis of the atmosphere of Jupiter. It was found to be composed of 86 per cent hydrogen and 13 per cent helium. The remaining 1 per cent consisted of traces of methane, ammonia and water vapour. The *Galileo* spacecraft dropped a probe into the planet, revealing wind speeds of 373 miles per hour (600 km/hr) and small concentrations of helium, neon, oxygen, carbon, water and sulphur. Although the sample was taken at only one point on Jupiter's vast surface, it is likely to be typical of the whole planet. Jupiter is one of the large planets known as gas giants. It has no solid surface, being composed of gas and liquid, although it has a rocky core.

Interesting Moons

The close-up views of the planets' satellites often proved to be just as exciting as the planets themselves. In the case of Jupiter, its larger moons Io, Europa, Ganymede and Callisto all have interesting features. Io was found to be extremely active volcanically, with molten lava and great plumes of sulphur ejecting from numerous volcanoes and rising to heights of up to 300 miles (500 km) above the surface. Europa by contrast was found to have an icy surface. The images showed brown streaks on the surface typically about 12.5 to 25 miles (20–40 km) wide and the *Galileo* spacecraft was able to see a cracked surface with the appearance of ice floes, suggesting evidence of liquid water beneath the ice.

Ganymede is the largest satellite of Jupiter and the largest in the solar system, with a diameter greater than that of Mercury. It has an iron-rich core with a permanent magnetic field, a rocky mantle with a thin atmosphere and an ocean of liquid water deep beneath the surface. Ganymede takes only 7.2 days to circle around Jupiter in a synchronous orbit. The probes were able to detect changing magnetic fields and electric currents near the surface, and the most likely cause of this would be an ocean of salt water. There are deep furrows in the icy crust and these can be explained in terms of a system of tectonic plates similar to the Earth's crust. The evidence for liquid water on Europa is even stronger and the possibility that both Ganymede and Europa could support some form of primitive life has created great excitement.

Callisto is the outermost of the true satellites of Jupiter, and it takes 16.7 days to complete an orbit. It has a thin atmosphere of nitrogen and carbon dioxide, and like Ganymede and Europa it may have liquid water under its icy surface. One of Callisto's most striking features is a set of concentric rings left behind by a huge impact millions, or perhaps billions, of years ago. The crater was named 'Valhalla' and the impact was so great that the outer rings are 1864 miles (3000 km) in diameter, and in some places huge spires of rock up to 100 metres (328 ft) in height have been thrown up. The source of heat to retain liquid water in such a cold part of space is probably created by radioactive decay inside the crust and the mantle of the moons. However, the surface of the planet is very cold; probes have measured 155 °K (–118 °C) at noon and 80 °K (–193 °C) at night. With the advantage gained by the use of space probes, scientists believe Jupiter has at least 63 satellites. This number will doubtless rise when more sensitive probes make the journey to Jupiter and the outer planets. Many of the satellites are not true moons but captured asteroids – identified by the fact that they are irregular ovals rather than true spheres.

Exploring Saturn

The planet Saturn has also revealed some of its secrets to the space probes *Voyager* and *Cassini*. The number of known satellites of Saturn is now 60, but only seven of these are large and spherical. New details of the ring system were revealed, showing the small 'shepherding' satellites called Prometheus and Pandora confining the particles on what is known as the F band to a fixed orbit. The rings consist mostly of lumps of ice, but some of the particles have a rocky core. The thickness of the rings proved to be a mere 10 metres (33 ft); they are not visible from the Earth when they are viewed edge on.

Saturn's largest moon is Titan. It was discovered by Christiaan Huygens (1629–95) in 1655. Titan is heavy enough to retain an atmosphere, of which 90 per cent is nitrogen, and the rest is mainly methane and other hydro-carbons. The *Cassini* spacecraft showed that the surface of Titan is partially liquid and free of craters, and that it is still undergoing dynamic changes. There has been much discussion about the possibility of Titan supporting life, due to the presence of complex organic molecules on its surface and in its atmosphere, but one of the main stumbling blocks is the very low surface temperature of only 95 degrees above absolute zero. For the necessary chemical reactions to create life, much higher temperatures are probably needed.

Exploring Uranus and Neptune

Probes have also visited Uranus and Neptune. In 1986 *Voyager 2* found Uranus to be a very featureless world, but the Hubble Space Telescope discovered a system of belts and zones. The rings of Uranus had been discovered in 1977 during an occultation (eclipsing) of a star by the planet. *Voyager* enabled a further study, showing them to be different from those around Jupiter and Saturn, and perhaps formed more recently. *Voyager* discovered ten additional small moons including some small ring-shepherding satellites similar to those of Saturn. The most interesting moon is Miranda; the *Voyager 2* image shows a world that underwent a shattering collision millions of years ago, Miranda has re-formed into a spherical shape, but with deep valleys and cliffs twice the height of Mount Everest.

The planet Neptune also has a ring system. The largest moon is Triton, discovered in 1846 and with a circular but retrograde motion around its planet. The tidal forces on Triton are so great that in the near future it could break into many small pieces – this would create a very spectacular ring around Neptune to rival those around Saturn.

Dwarf Planets

For many years after its discovery in 1930 by Clyde Tombaugh (1906–97), Pluto was considered to be the most distant planet in the solar system. But in 2006 a formal decision was made to downgrade this icy little world with a diameter of 1,490 miles (2,390 km) and a mass of about 2 per cent of that of the Earth from the status of a main planet to that of a dwarf planet. This decision was made in the light of the identification of several new solar system bodies similar in density, orbit and size to Pluto. The largest of these, Eris, was discovered in 2005 by a team led by the American astronomer Michael Brown (b. 1965). With an estimated diameter of at least 1,550 miles (2500 km), it is larger than Pluto. Ceres, the largest of the asteroids in the belt that lies between Mars and Jupiter, is also now classified as a dwarf planet. None of these dwarf planets has yet been visited by any space probe, but this will be rectified in 2015, when the *Dawn Mission* visits Ceres, and the *New Horizons* probe reaches Pluto.

One of Pluto's three moons, called Charon, has a diameter of nearly 1243 miles (2000 km) and is only about 12,426 miles (20,000 km) away, so it may claim to be a double planetary system. Pluto's orbit passes inside the orbit of Neptune, and it takes 248 Earth years for Pluto to orbit around the Sun.

The Kuiper Belt and the Oort Cloud

Beyond the orbits of Neptune and Pluto are many chunks of rock and ice, numbering several billions. These chunks of rock and ice are called comets. Some of them are located in a ringed area called the Kuiper Belt and others are found in an approximately spherical region called the Oort Cloud. When the orbit of one of these comets takes it near the Sun, the Sun's heat melts the ice, and as it evaporates it forms the familiar spectacular 'tail' sometimes visible to the naked eye. The Kuiper Belt was named after the American astronomer Gerard Kuiper (1905–73), who proposed the existence of the belt in 1951. It is believed that there are about 200 million comets in the Kuiper Belt. The Oort Cloud, where most comets are thought to exist, was discovered by the Dutch astronomer Jan Oort (1900–92) in 1950.

New Telescopes and New Techniques

The space age has also seen the development of instruments that can be used closer to home but which can be employed to study objects in deep space. In 1990 the space shuttle *Discovery* launched the Hubble Space Telescope (HST). It was a successful launch, but soon afterwards there was great dismay when it was discovered that the 2.4-metre (7.8 ft) objective mirror was flawed and a haze surrounded all of the star images. It was three years before the defect could be corrected, but the mirror was successfully upgraded in 1993 and the telescope was enhanced again in 2002. After the second upgrade the resolution of the telescope was a tenth of a second of arc. This meant that the telescope had the power to see something the size of a 1-centimetre (0.4 in) diameter coin at a distance of 12.4 miles (20 km). The HST is due for a fifth and final upgrade that is planned to extend its life by several more years.

Throughout the history of astronomy, new instruments have led to new discoveries. The same was true for the HST. Free from the Earth's restricting atmosphere and with clear views in all directions over a wide range of the optical spectrum, the HST was quickly making new discoveries. More detail was seen on planets, and at the limits of observation sharper images were seen of proto-planetary systems around other stars, star clusters, nebulae, galaxies and quasars. The whole sky came under scrutiny, and when other specialized telescopes joined Hubble it was mapped in infrared and ultraviolet wavelengths. In spite of the great cost of putting a telescope into orbit around the Earth, the HST was seen as one way forward for astronomy and cosmology. For centuries the twinkling of the stars, caused by the Earth's atmosphere, restricted the sharpness of the images observed with earthbound telescopes. And for centuries it was assumed that there was no solution to the problem. Then, in parallel with the development of space observatories, along came the advanced technologies of active optics and adaptive optics applied to ground-based observation. The former technique applies computer technology to adjust the mirror every few seconds according to changes in temperature and to keep the focus of the mirror sharp. The latter uses sensors to follow the observed twinkling of the stars so that the software can minutely readjust the shape and direction of the mirror to correct the variation. At present the ground-based telescopes can resolve to about 0.3 arc seconds, three times coarser than that of the HST, but because they are built on the ground they can be constructed much larger and more cheaply than the HST, and it is only a question of time before they are producing sharper images.

There is a technique used in exploring the heavens called interferometry. This is a method of enhancing the resolution by combining the electromagnetic radiation detected by two or more telescopes. It has been used in the radio waveband for many years, but it is now being applied to shorter-wavelength optical observations taken by telescopes such as the twin Keck Telescopes on Mauna Kea in Hawaii. Each of these telescopes has an array of 36 hexagonal mirrors, all independently moveable, and the combined total is equivalent to a telescope with an objective mirror of 85 metres (279 ft).

The HST was just the first of NASA's 'Great Observatories' in space. It was followed in 1991 by the Compton Telescope which detected hard X-rays and gamma rays from space, and the Chandra Observatory in 1999. These telescopes detect photons from the very highest frequencies of the electromagnetic spectrum. Light at these frequencies is unable to penetrate the Earth's atmosphere and, therefore, telescopes for detecting them can only operate from above the atmosphere. When X-rays strike metallic surfaces they tend to penetrate them, unless they strike at a very shallow angle in which case they are reflected. Special telescopes have been designed to focus X-rays using concentric nested paraboloid and hyperboloid mirrors, and much of the sky has now been mapped at these frequencies. Gamma rays are even more difficult to focus. They can, however, be controlled using crystals and tiny directional holes called collimators. Fortunately the gamma rays have very high energy levels and they are easy to detect.

The last of the Great Observatories is the Spitzer Space Telescope, launched in 2003, which maps the sky at infrared frequencies. It studies the light from planets, comets and interstellar dust clouds in the infrared part of the spectrum. The wavelength coverage of the space telescopes is augmented by two observatories that detect ultraviolet photons (the Extreme Ultraviolet Explorer [EUVE] and the Far Ultraviolet Spectroscopic Explorer [FUSE]), both launched in the 1990s.

SOHO – The Solar and Heliospheric Observatory

The Sun is the brightest object in the sky and although we know a great deal about it there is still much to learn by studying it from space. The Solar and Heliospheric Observatory spacecraft, or SOHO, was built by a consortium of 14 European countries and it was launched in December 1995 to study the Sun. It has been able to plot temperatures and convection currents inside the Sun at temperatures of up to one million degrees Celsius. It can see right into the core of the Sun where nuclear fusion of hydrogen into helium is taking place. Originally the mission was only expected to last for two years, but such has been its success that it has been continually extended to 11 years so that a complete sunspot cycle can be studied.

From its orbit high in space SOHO has also studied the Sun's surface and phenomena such as the solar wind, the stream of particles (mainly electrons and protons) that emanate from the Sun. SOHO has also discovered over 1300 new comets. A few of them have elliptical orbits similar to Halley's Comet, but the majority travel far into space without returning.

The Chandra X-ray Telescope

The Chandra X-ray Telescope has the distinction of being the heaviest payload ever launched by a space shuttle. It is named after the Indian-American Nobel prizewinner and astrophysicist Subrahmanyan Chandrasekhar (1930–95), who was the first to recognize that there was an upper limit to the mass of a white dwarf star. The Chandra X-ray Telescope was put into an elliptical orbit around the Earth in 1999 and it is still sending back valuable information about the X-ray universe. The X-ray universe provides more information about high-energy particles in space, which can originate from gases at multi-million degree temperatures; or from regions with intense magnetic and gravitational fields, such as occur very close to a black hole. X-rays are well outside the visible spectrum; they need special grazing incident mirrors for focusing, and the images are reproduced in false colours to enhance salient features.

Deep Impact

The USA celebrated Independence Day 2005 by making the first contact between an artificial object and a comet. The spacecraft *Deep Impact*, with a mass about the size of a small car, struck the comet Tempel 1 on 4 July and the event was observed and photographed by many telescopes. The speed of the impact was 25,000 miles per hour (40,234 km/hr). From the impact data, astronomers were able to make deductions about the nature of the comet's surface, its mass and its chemical composition.

The Role of the Amateur Astronomer

When we hear of the latest developments in astronomy and cosmology it looks increasingly as though the only exciting discoveries still to be made are with the use of space probes and orbiting telescopes or high-budget earth-bound telescopes. There are, however, still opportunities for the amateur astronomer, especially since even quite sophisticated telescopes and detectors can be purchased relatively cheaply.

Amateur astronomers play an important role in the detection of both supernovae and comets, both of which are discovered by painstakingly charting the sky and looking for changes in the Milky Way and local galaxies. For example, one amateur, the Reverend Robert Owen Evans (b.1937), has more supernovae to his name than any other astronomer, and it was the amateur David Levy (b.1948) who co-discovered the comet Shoemaker-Levy on its course to impact with Jupiter.

21

THE BIG BANG AND
THE CREATION OF
THE UNIVERSE

In 1927 the Belgian astrophysicist Georges Lemaitre (1894–1966)
discovered a particular result from Einstein's equations of general
relativity that suggested the universe could be expanding. A similar
result had been obtained by the Russian Alexander Friedmann
(1888–1925) in 1922, but had been largely ignored by the astro-
nomical world. Lemaitre, however, went on to suggest that the
universe must have had a starting point – in other words, the
Big Bang – when the whole of space–time, and all the matter
and energy within it, were created in a single instant.

In the late 1920s George Gamow (1904–68) correctly
suggested that the stars were powered by nuclear fusion;
the temperatures were high enough to create helium atoms
from hydrogen atoms with the release of vast amounts
of energy. After the Second World War many new theories

and predictions about the universe were made, many of which were based around the idea of a Big Bang, and in the 1950s George Gamow became the leading proponent of the idea. He predicted the presence of background radiation and he made a good estimate of its temperature. He and others were able to work out many details of this theory of the creation. In particular, they suggested that many chemical elements had to be created during an early, hot and dense period of the universe.

Looking for the Evidence

As we have already seen, the Big Bang was not the only theory for the origin of the universe, the chief contender being the steady-state theory. The main challenge for this theory was to explain Hubble's observation that all the galaxies seemed to be rushing away from each other. The steady-state proponents accounted for this by suggesting the continual creation of a few atoms per year in every few cubic miles of space. Whilst this did require matter to be formed out of nothing, they maintained this was far less of a problem than the contention that the Big Bang created everything in a single instant of time. If the universe really had started with a Big Bang, Gamow and his co-workers argued that the very high temperatures shortly afterwards meant that space would have been saturated with radiation. Later, as the universe cooled, matter would begin to dominate, but even several billion years later, the early thermal radiation would still be present. They even calculated that by now it should have cooled to a temperature of about 5 degrees above absolute zero, and thus it should be observable in the radio waveband. The radiation is known as the cosmic microwave background (CMB).

It was not until the 1960s that anyone made a systematic search for the CMB. Even while Robert Dicke (1916–97) and his colleagues at Princeton University were designing a microwave antenna for this purpose, the CMB had already been discovered by accident a few miles away. In 1965 two employees of the Bell Telephone Company, Arno Penzias (b. 1933) and Robert Wilson (b. 1936) were using a sophisticated horn antenna to track communication satellites and to pick up radio messages. They had encountered a problem, however. They found that in whichever direction they turned their antenna they would always pick up a background noise, and try as they might they just could not get rid of it. Discussion with the Princeton astronomers led them to realize that they had stumbled across the diffuse background radiation by accident, and subsequent observations by Dicke and his team confirmed the discovery. Their observations showed that the radiation had a thermal spectrum, with a temperature only a couple of degrees lower than that predicted by Gamow.

The discovery of the CMB gave a tremendous boost to the Big Bang theory, and sounded the final death knell for the steady-state theory. In a relatively short time some advanced theories were put forward to explain the first few moments of the universe. The Big Bang model evolved as scientists applied their minds to the problem. The conditions of temperature and density at the time of the Big Bang were unimaginably high, but by the 1980s some very sophisticated theories were available and cosmologists were extrapolating their ideas right back to the very first instant of creation.

The Cosmic Microwave Background

Far more detailed measurements of the properties of the CMB were undertaken in the later part of the 20th century using sophisticated equipment and techniques. Due to the way in which water in the atmosphere absorbs radiation, most of the experiments have been carried out by balloons and satellites at very high altitude. In particular the Cosmic Background Explorer launched in 1990 established the temperature of the CMB as only 2.7 degrees above absolute zero. The radiation is observed to be close to uniform across the sky; the Wilkinson Microwave Anisotropy Probe has limited any deviations in temperature to less than one part in over 100,000. This limits any fluctuations in the distribution of matter and energy in the universe at the time that the radiation was originally emitted, thus giving astronomers one of the only observational constraints in the earliest epochs of the universe.

A New Scale of Measurement

In the 19th century the German physicist Max Planck (1858–1947) devised a form of measurement that we now call the Planck scale in an attempt to simplify the equations of atomic physics. In the 20th century the idea was extended to simplify the values of universal constants such as the speed of light, the gravitational constant and the unit of charge. The growth of nuclear physics and quantum mechanics showed a need for a system of very small units to deal with the microcosmic world of the atom. Thus, when the theory of black holes developed, the Planck mass was defined as the mass of a black hole with a Schwarzschild radius the same order as the Compton wavelength used in quantum mechanics. It therefore needs some knowledge of black holes and quantum mechanics to understand the definition. The Planck mass works out at 2.18×10^{-8} kg. It is just imaginable as the mass of a barely visible flea. The corresponding Planck unit of length is 1.62×10^{-35} metres, unimaginably tiny even compared to an electron. The Planck time is the time it takes light to travel one Planck distance: 5.39×10^{-44} seconds. Both of these units can be regarded as effectively the smallest possible components of length and time; we cannot conceive of a smaller time interval than the Planck time or a shorter distance than the Planck distance. The units are important when we study the early phases of the Big Bang, at which time the whole universe was unimaginably small. In describing the evolution of the universe we need to deal with time intervals down to 10^{-44} seconds at the creation and time intervals of several billion years as we reach our present time. A linear scale could not possibly cope with the range of times, temperatures and distances involved in the unfolding of the story of creation.

A Journey Back in Time

The observational starting point for cosmologists is the expanding universe of galaxies receding rapidly from us. Let us imagine we reverse time and run events backwards so that all the galaxies are rushing towards each other instead. It is possible in our time-reversing universe to work out how far apart they were a billion years ago. The density of matter in the universe would have been higher at that time, but the galaxies would still be very distant. However, if we keep proceeding backwards in time in steps of a billion years we reach a point, long before the evolution of life on Earth, when all the galaxies were so much nearer to each other. Close to 13.7 billion years ago, there comes a point where all the matter and energy in the universe is compressed into a tiny space. The resulting temperature is so high that most of the mass of the universe is in the form of radiation. We cannot see and measure this early universe and must rely on our understanding of the laws of physics and mathematics to determine what must have happened in the earliest phases. Even so, our understanding is severely constrained by our inability to combine relativistic and quantum physics – which is essential if we are to fathom how infinitely small yet massive concentrations of matter and energy behave.

Unifying the Forces of the Universe

Before tackling the Big Bang we need to understand the concept of force. Force is one of the keys to the nature of matter. As babies, we soon acquire a knowledge of the force of gravity. We discover the force of magnetism when we play with a magnet, and we discover electricity in the form of electrostatic force when we find that after pulling a plastic comb through our hair it can pick up small particles. In addition to these forces there are two others, called strong and weak nuclear force, but they act over a very limited range and can only be studied in extreme conditions. Physicists have consistently tried to find a theory that unifies all the known forces acting in the universe into a single force. Electric and magnetic forces were once thought to be distinct from each other but, as described in an earlier chapter, they were unified by the work of James Clerk Maxwell (1831–79) in the 19th century. In the 20th century Albert Einstein (1879–1955) tried to unify electromagnetic force with gravitation, but his efforts met with little success.

Weak nuclear force was needed to explain some of the observed aspects of radioactivity. It acts on sub-atomic particles such as quarks, electrons and neutrinos. Experiments at CERN in the 1980s created collisions at such high temperatures that it was impossible to tell the difference between weak nuclear force and electromagnetic force. Weak nuclear force was shown to be a special case of electromagnetic force at certain high temperatures, and these two forces have now been successfully unified. Thus the number of forces in the universe comes down to only three: gravitational force, electroweak force and strong nuclear force. Extending the idea, physicists theorize that at even higher temperatures – such as in the initial stages of the Big Bang – all three forces would be indistinguishable. It is likely that Einstein's dream of a unified field theory was a reality in the earliest stage of creation. Physicists hope to be able to reconcile all three forces into a single force. A new particle accelerator, the Large Hadron Collider, will smash atomic particles together at energies approaching those present soon after the Big Bang. It is hoped the experiments will help us to understand how all the forces could have been unified, and how they might have behaved in the first era of the Big Bang.

We shall now go right back to the moment of the creation and examine the presumed behaviour of the universe from the first instant. We can say nothing about what happened before the Big Bang or how it was caused; such questions are probably more of a philosophical or a theological nature.

Back to the Beginning of Time

Let us turn back the clock and fix on an instant of time that we shall define as zero. We call this the Planck Era, and it lasts from time zero to 5×10^{-44} seconds, or one Planck unit of time. At this point, the universe would have fitted into the nucleus of an atom several billion times over. However, our limited knowledge means that there is much speculation involved in trying to envisage anything detailed about the universe at this time. We assume that it was so hot that even at the end of the era the temperature was of the order of 10^{32} kelvin, and in such hot and dense conditions, there must have existed some very exotic laws of physics. It is probable that all four forces were indistinguishable. As the universe expands out from the point of creation, it cools. At some point gravity separated or 'froze out' from the other forces, heralding the second era of expansion.

This era is called the Grand Unification Epoch (GUE), since at temperatures higher than 10^{27} kelvin, it was sufficiently hot for all the forces other than gravity to remain unified. During this period most of the universe's energy was in the form of radiation, but under the laws of quantum mechanics pairs of elementary particles were frequently created by the process known as pair production. Particles of both matter and antimatter were created and there was a great conflict as they annihilated each other whenever they came into contact. The battle was won in favour of matter. If the forces of antimatter had won then a very different universe would have been created. The GUE lasted until about 10^{-35} seconds, at which point the universe cooled to the point where the strong nuclear force 'froze out'. Although the GUE lasted for a very brief period of time indeed, it was still a hundred million times longer than the Planck Era that preceded it.

More Rapid Changes

Next, the universe is thought to have entered a phase called the Inflationary Epoch in which it underwent an exponential expansion, growing to about 1050 times its previous size in under 10^{-33} seconds. Originally mooted by Alan Guth (b.1947) in 1981, the concept of a period of inflationary evolution accounts for some otherwise puzzling properties of the CMB, although we have no direct evidence for this phase. First, the spectrum of the CMB is completely uniform in temperature, despite the enormous physical separation between different sides of the present-day universe. During the Planck Era, the universe was like a small, hot and homogenous 'soup', where every bit was in contact with every other bit, and reached a uniform temperature. Subsequent inflation then instantaneously dispersed all the matter outward in all directions to give uniformity.

The tiny temperature fluctuations of the CMB tell us that any deviations in density in the initial universe were miniscule; yet it is these slight overdensities that provide the starting points for the condensations of matter under gravity. Inflation provides a mechanism whereby such tiny fluctuations are stretched and magnified to the physical length scales of the galaxies they will later form. Finally, the angular scale of the CMB fluctuations suggests that the present-day universe has a 'flat' or Euclidean geometry (where 'flat' here does not have its normal linguistic sense, but is used to imply the kind of familiar geometry where parallel lines only meet at infinity, and the internal angles of a triangle still add to 180°). There would not normally be any reason to suppose that this would be the eventual geometry, unless an inflationary period stretched and flattened the universe, regardless of its initial shape.

The Quark Era

After this period of exponential growth, the universe was dominated by quarks and antiquarks. Quarks can be seen as the building blocks of all the elementary particles. The collisions and reactions between atomic particles, and the properties of the particles, can be explained more simply in terms of quark theory.

The first part of this era, until electromagnetic and weak forces separated out when the universe was 10^{-12} seconds old, is often called the Electroweak Era. This left only electrostatic and magnetic forces united, and we know they remained united because they are the same force when met at what we would consider to be 'room temperatures'. The universe was still very hot and energetic, and even as new particles were created they were destroyed and changed into high-energy photons on meeting an anti-particle. Hence radiation was still dominant over matter. The quarks were at first unable to combine with each other to form the heavier atomic particles, as the temperatures were too high. As the age of the universe approached the end of the first microsecond it cooled sufficiently for quarks to do their work. The universe entered the Hadron and Lepton Era, when atomic particles such as protons, neutrons and other baryons were created.

At last it was time for the more familiar laws of physics to rule the universe, but there were still many interchanges between matter and energy. These quantities are related by Einstein's law $E = mc^2$. The universe came of age by the time a whole second had passed since the moment of creation. The temperature had fallen to a mere million or so degrees and the radius of the universe had expanded to about 187,000 miles (300,000 km) – roughly three-quarters the distance of the Moon from the Earth.

The expansion of the universe continued. There were plenty of protons and electrons around to form atoms of hydrogen, but there were even more high-energy protons that frequently collided with the newly formed atoms and caused them to split apart. It was not until temperatures fell below a million degrees that the hydrogen atoms became more stable, and then a few atoms of helium were formed.

The Recombination Era

As the universe expanded the density fell rapidly until matter was able to dominate. Particles were stable and they existed for a longer period of time, in particular the atoms of hydrogen and helium. Up to this point, the particles and photons had been tied together, continually interacting with each other. But as the particles recombined to form stable atoms, the photons became free, and many streamed towards us. The moment when the photons were 'last scattered' by the matter marks the first observable feature in our nascent universe – the CMB. We will never be able to observe the earlier universe with our telescopes; we can only conjecture what happened before the CMB from what we know of the laws of physics and mathematics.

About one billion years after the Big Bang, protogalaxies and protostars began to form. The most massive stars lived only a few million years, evolving into supernova explosions, and scattering the heavy elements they had fused in their core across interstellar space. Subsequent cycles of star formation and evolution continued to enrich their surroundings, and stars formed later in the lifetime of galaxies resembled our Sun. Some of these stars formed planetary systems, and possibly other Earth-like planets were created. At least one of these planets evolved in a fascinating way. It developed life.

22

DARK MATTER AND DARK ENERGY

Dark matter and dark energy together make up some 96 per cent of all the mass and energy in the universe. But their nature remains a mystery. Dark matter was proposed decades ago to explain why galaxies hold together, while dark energy is a more recent prediction which explains why the universe is not just expanding, but is doing so at an ever-increasing pace.

Ever since the dawn of astronomy observers have wanted to know the distance to the stars and the scale of the universe. The methods by which these distances have been measured since the early days of astronomy provide a little history of its own.

Measuring the Universe

The distance to objects in our solar system can be measured by the method of parallax, whereby two images of the same object can be viewed from two places a known distance apart. The Moon, for example, can be seen at the same time against a different star background from two widely spaced points on the surface of the Earth. The difference in its position against the distant stars can be used to calculate its distance from Earth.

Measuring a parallax for the planets is more difficult, but the method still provides a way of finding their distance from the Earth. Measuring a parallax for the stars was a far more difficult problem. It took centuries to solve, but with modern detection techniques the distance of stars about 100 light years away can also be measured by the method of parallax; the baseline used for the stars is a diameter of the Earth's orbit.

In the 20th century, methods to calculate longer distances became available. The first was developed by Henry Norris Russell (1877–1957), who with Ejnar Hertzsprung (1873–1967) discovered the relationship between the magnitudes and spectra of stars. Confusingly called the 'spectroscopic parallax' method, it has little to do with trigonometry, relying instead on the observed properties of stars. The H-R diagram provides the correlation between a star's spectral type and its absolute magnitude, by which we mean its magnitude if it was located at the 'standard' distance of 10 parsecs from the Earth. Using the distance–magnitude relationship, the star's distance can be inferred from its observed brightness. While less precise than trigonometric parallax, this technique enabled astronomers to estimate distances as far away as 10 kpc.

To measure distances beyond 1 kpc and as far away as 30 Mpc, the best technique is the one developed by Arthur Eddington (1882–1984) and Henrietta Leavitt (1868–1921) using the Cepheid variables. The Cepheids are variable stars glowing brighter and dimmer over a period of several weeks or days. They have a well-known period–luminosity relationship; thus if we know the period of variability of a Cepheid then we know its absolute luminosity. By comparing this with the luminosity observed from Earth we can calculate its distance. Cepheids separate into two types, commonly known as population I and population II, according to their luminosity. Both are useful distance indicators, but the brighter population I Cepheids cover a much wider range of distance and we can observe them in our neighbouring galaxies, giving us indicators to estimate the distance to the nearer galaxies.

Other techniques have been developed to measure galactic distances. In the 1970s Brent Tully and Richard Fisher discovered a relationship between the properties of the 21 cm hydrogen line observed in the radio spectrum of a spiral galaxy and its intrinsic luminosity of the galaxy. The hydrogen line is emitted from cold clouds of gas that lie between the stars in the disc of a spiral galaxy and which, like the stars, rotate around the centre. This rotational motion is detectable as a broadening of the 21 cm absorption line due to the Doppler effect, and the amount of rotation (and hence the observed width of the line) is driven by the internal gravity of the galaxy. The broader the line, the higher the galactic mass driving the motions, and thus the larger the intrinsic luminosity of the galaxy. It was possible from this broadening effect to calculate the absolute magnitude of a galaxy at the standard 10 kpc, and by comparing the absolute magnitude to the observed magnitude the distance of the galaxy could again be estimated. The Tully–Fisher technique can be used to measure distances up to 150 Mpc.

There is another very useful technique that can be used for measuring distant galaxies. It is based on the observation of the rare event called a supernova Type 1a. The supernovae are so bright that they can be seen in galaxies at distances of well over 1000 Mpc. These supernovae are due to a white dwarf in a binary system accreting matter from its secondary until it is at the Chandrasekhar limit of around 1.4 solar masses for a catastrophic core collapse in a supernova explosion. As the same mass is involved in each explosion, the event follows a similar pattern in terms of the intrinsic luminosity of the outburst and how this is related to the rate at which this brightness subsequently fades away. Thus the observations of the apparent magnitude of a super-nova explosion and the rate at which it fades enable a calculation to be made of its distance.

All of these distance-measuring techniques neatly overlap with each other so that a three-dimensional image of the universe is gradually being built up by astronomers. However, there is still one important technique that needs to be discussed; this is the redshift of the distant galaxies, and it is of great significance for it provides the key to the age of the universe.

Hubble's Constant (H_0)

During the 1920s Edwin Hubble (1889–1953) and Milton Humason (1891–1972) recorded the spectra of many galaxies using the 2.5-metre (100 in) telescope at Mount Wilson. As early as 1917 Vesto Slipher (1875–1969), working at the Lowell Observatory in Arizona, discovered that the spectra from the galaxies were noticeably shifted towards the red end of the spectrum. He rightly concluded that the galaxies were moving away from us. Hubble, however, used the Cepheid variables technique to estimate these distances and derive the distance–redshift relationship known as Hubble's law. This relationship can also be used to derive an estimated distance of a galaxy from its much more easily observed spectral redshift.

During the 20th century, a great deal of effort and research were devoted to an accurate determination of Hubble's constant, culminating in a key project using the Hubble Space Telescope. The best estimate for Hubble's constant is now generally taken as their result of 70 +/- 7 km/s/Mpc, published in 2001. The significance of this result lies in the fact that the inverse of Hubble's constant gives an estimate for the age of the entire universe.

Consider a distant galaxy rushing away from us. If we know its distance from us and its speed of recession, then the time it has taken to separate from us and the rest of the universe after the Big Bang is given by:

Time = (the distance from us) / (its speed away from us)

As Hubble's constant is simply the averaged ratio of speed/distance determined from a large number of galaxies, its inverse yields the required time. If $1/(H_0) = (1/70)$ seconds Mpc/km, and using the unit conversion factors 1 Mpc = 3.09×10^{19} km and 1 year = 3.156×10^7 s, we can estimate an age of the universe:

$1 / (H_0) = (1/70) \times (3.09 \times 10^{19}) / (3.156 \times 10^7)$ years

$= 1.4 \times 10^{10}$ years (or 14 billion years).

Hubble's constant thus provides a good first approximation for the age of the universe. However, as we shall see later in this chapter, the study of the expansion of our universe was going to throw up some surprises for astronomers.

Using the 21 cm Hydrogen Line

We have seen how the spiral structure of our own galaxy, the Milky Way, has been discovered. The Sun lies well out in one spiral arm of the galaxy. At the centre is a great bulge with the spiral arms lying in a disc around it. The bulge is surrounded by a halo of globular clusters consisting of old red stars, but the spiral arms contain younger blue-coloured stars. Between the young stars in the disc of the galaxy are clouds of cold, molecular gas and dust which are the reservoir from which stars form. The existence of the dust had long been known from the dark patches apparent within the Milky Way, showing how the dust clouds can be so dense that they completely obscure the light from the stars behind them. The presence of vast quantities of gas was revealed from observations of the luminous nebulae surrounding clusters of newly formed young blue stars. The energetic ultraviolet light from the stars heated and ionized the gas atoms, which produced their own light when they later recombined, causing the nebulae to shine. Analysis of the emission from the gas enabled astronomers to measure the temperatures and densities of the gas nebulae; much of the radiation comes from 'forbidden' transitions of ions of common elements that can never be observed in the laboratory since they require an extraordinary low density to occur.

Away from regions of active star formation, the gas clouds remain cold, and thus hidden to optical observations as they are either neutral or even molecular. The discovery of the extent of the neutral hydrogen in our galaxy and others had to await the development of radio astronomy after the Second World War. During the 1940s the theory of the 21 cm hydrogen line was developed. This is a spectral feature in the radio waveband that is generated by what is called a 'spin flip' transition of electrons in hydrogen atoms. This is a spontaneous event that occurs very rarely – an individual atom may undergo such a change only once every ten million years – but it is still detectable because of the vast quantity of hydrogen in the galaxy. As radio astronomy developed radio emission lines from more complex molecules were discovered.

Observations of the 21 cm line were used to map out the density and distribution of gas in the plane of the galaxy at much further distances than could be obtained by studying the distribution of the stars. It is much easier to measure the Doppler shifts from the 21 cm line of different clouds of hydrogen than by amassing the shifts from thousands of individual stars, so the radio observations were also of major importance for determining the rate at which our galaxy rotates, and how this changes with distance from the centre.

Rotation of the Galaxy and the Missing Matter

As we saw with the Tully–Fisher distance indicator, the rate at which a spiral galaxy rotates is directly due to its gravitational mass. Any individual object in orbit around the centre of a galaxy – such as a neutral gas cloud, nebula or star – is responding to the gravity of all the mass at smaller radii. Thus by plotting the way the rotational velocity of objects in a galaxy changes with radius (known as a 'rotation curve'), astronomers knew they could estimate the entire mass of a galaxy, and how it was distributed. Early attempts to determine the rotational velocity of galaxies used the Doppler shift observed from the absorption lines in stellar spectra, or the narrow emission lines of nebulae. By the late 1970s Vera Rubin (b. 1928) and her colleagues established a problem with the observed rotation curves of galaxies, which was later confirmed by more comprehensive observations using the radio 21 cm line as a tracer of the internal galactic dynamics.

All the results showed that the outer parts of the disc of all spiral galaxies (including our own) were rotating much faster than expected. They had sufficient speed to escape completely from the galaxy's gravity, but they remained attached. The only explanation was that there was far more gravitational mass in the galaxy than was indicated simply by the amount of stars and gas directly observed. A large quantity of invisible matter was required that did not give off radiation at any wavelength; and in all spiral galaxies the amount of such 'dark matter' was estimated to be greater than the visible mass by a factor of ten. But what was all this invisible matter?

The idea of 'missing matter' was not a new concept, as it had been discovered as early as the 1930s when Fritz Zwicky (1898–1974) was studying clusters of galaxies. The motions of individual galaxies within a cluster are due to the gravitational attraction of the galaxy to the mass of the rest of the cluster. Zwicky was able to show that again, the galaxies were moving too fast to be responding simply to the gravity of the visible galaxies in a cluster. There had to be much more mass present, but this time it outweighed the observable components by a factor of over 100. Zwicky's original results have long since been confirmed for many clusters of galaxies, from observations of the internal dynamics of a cluster and also from studies of gravitational lensing.

Dark Matter

Although it is well established that most of the gravitational mass in the universe is invisible at all wavelengths, the nature of the dark matter remains an open question. Possibilities range from 'ordinary' matter which is comparatively well understood, such as brown dwarfs (failed stars), large planets, neutron stars or black holes; to far more exotic (and hitherto undiscovered) subatomic particles such as axions or new types of neutrinos. The latter explanation proposes that a better understanding of particle physics is fundamental to explaining the motions of entire galaxies and clusters of galaxies.

Dark Matter and Gravitational Lensing

Einstein's theories reinvented our interpretation of gravity not so much as Newton's 'force at a distance', but as the way in which space and time warp around the location of a massive object. An important consequence is that light travelling through space will sometimes find itself following a warped path, and indeed the observations of the deflections in the positions of stars whose light passed near the Sun during the 1919 total solar eclipse was the first observational confirmation of Einstein's theory. Today astronomers observe much more complicated distortions of the light as it passes by a large mass on its journey to Earth. Galaxies that lie behind (and at a far greater distance than) a cluster of galaxies sometimes have their image both magnified and greatly distorted as their light passes through the cluster – this is known as gravitational lensing. Often this distortion takes the form of multiple arcs and arclets. However, the amount of distortion traces the total gravitational rather than the visible mass, so a detailed mapping of the positions of such gravitational mirages can reveal the presence and distribution of dark matter within a cluster.

Clustering of the Galaxies

Since the first studies of galaxies in the early part of the 20th century, it was obvious that the galaxies were not distributed uniformly on the sky – even when located well away from the obscuring effects of our Milky Way. This 'clustering' of galaxies was first properly quantified in the 1950s by George Abell (1927–83), who created an extensive catalogue of clusters from a detailed visual examination of photographic plates of the sky. His work also demonstrated that there was a range in cluster properties – not just in the number of galaxies, but also the shape and physical size of a cluster.

Redshift Surveys

Abell's work – and that of other astronomers in the mid-20th century – was limited to the study of a projection of the sky onto two dimensions. Even so, it was clear that the distribution of clusters was also non-uniform in the sky, with regions where clusters themselves seemed to form immense structures known as 'superclusters'. A full mapping of the true three-dimensional structure of the universe, however, involves the knowledge of the distance to all the galaxies. This is done most economically by estimating the distance from a measured redshift via Hubble's law; even so, the determination of the redshift of a sufficient number of galaxies is a huge observational undertaking. For this reason, early surveys were necessarily limited to only small regions of the sky. One of the first attempts at a comprehensive redshift survey was begun by John Huchra (b.1948) and Margaret Geller (b.1947) in the 1980s, and it eventually grew to include over 14,000 galaxies. The resulting map showed some amazing structures, including the discovery of what became known as the Great Wall – a broad filament of clusters and galaxies about 200 million light years distant, which extends over 500 million light years long.

This was just the first (and not the largest) of many such walls and filaments now known to permeate the entire large-scale structure of the universe, and which surround regions of empty space of a similar size, known as 'voids'. Today such surveys map out this cellular pattern of structure right across the universe.

Supermassive Black Holes

The presence of dark matter was discovered by observing the way motions of astronomical objects are dictated by the gravity of an invisible mass. In the same way, the orbital motions of stars and gas at the very core of many galaxies (both spiral and elliptical) have revealed the presence of immense masses at the centre. Our own galaxy is thought to have a dark object at its centre with a mass of around 2.5 million solar masses but contained in a region less than 20 light days across. Such supermassive black holes are found with masses up to several billion solar masses, and they are thought to lie dormant at the centre of nearly all massive galaxies. A galaxy where the central supermassive black hole is still accreting matter is radically changed in appearance, having a very bright core, and is known as an 'active' galaxy.

What Happens Next?

By the latter part of the 20th century, astronomers had thus gained a much better understanding of the mass content (both visible and invisible), the structure and extent, and the rate of expansion of the universe. It was only natural to return to the fundamental question that had been asked many times before. Will the universe last forever or will it stop expanding and collapse into itself? For many years cosmologists debated about whether the universe is 'closed', so that it will eventually fall back on itself and end in a 'Big Crunch', or whether it is 'open' and will continue expanding for all time at an ever increasing rate. The outcome depends on the mass content of the universe, and thus whether or not there is enough mutual gravitational attraction eventually to pull everything back together.

A further option – favoured by most astronomers because of the observed isotropy and homogeneity of the cosmic microwave background – was the 'flat' universe; one which lies on the boundary between the open and closed universe.

Consequently, in the 1980s and 1990s, ambitious programmes tried to measure the expansion of the universe more precisely than before, with the aim of determining its eventual fate. By attempting to extend Hubble's law to far further galaxies than had been possible previously, astronomers aimed to see if the relationship between velocity and distance still held, or whether it indicated the first stages of a gradual slowing, or deceleration, that would result from a flat or closed universe. The research was undertaken by two teams of astronomers, one led by Saul Perlmutter (b.1959) at Berkeley and the other led by Brian Schmidt (b.1967), then at Harvard Observatory. Both used observations of Type 1a supernovae to establish the distance to very early galaxies, and they independently discovered an astounding result. The expansion of the universe was not found to be continuing at a constant rate, nor was it beginning to slow down. Astronomers were amazed to find that the rate of expansion was increasing with time – it had been accelerating for the last six billion years! This result, announced in 1998, changed the face of modern cosmology, and has since been confirmed by many more and different observations to extend Hubble's law using different ways of estimating distances.

Dark Energy

In order to begin to understand why this unexpected acceleration had occurred, it was necessary to introduce a new concept which went by the (somewhat misleading) name of 'dark energy'. Einstein's famous equation $E = mc^2$ shows how matter and energy are related, and in some sense can be considered as equivalent entities. Astronomers were astounded to discover that to produce the accelerated expansion, dark energy had to account for around three-quarters of the entire mass–energy density of the universe. But a true understanding of dark energy remains elusive, although there are a number of current ideas.

For centuries it has been known that gravity is an attractive force; unlike electricity and magnetism it is not possible under normal conditions for gravity to be repulsive. However, suppose the fundamental nature of gravity changed from an attractive to a repulsive force on the very largest (and truly astronomical) distance scales. Then there might come a point when the galaxies drew a sufficiently immense distance apart that they began to be pushed away faster and further by gravity. Such ideas echo Einstein's original, but misguided, attempt to include in his equations a cosmological constant representing an anti-gravity that prevented the universe from evolving. It is possible that a modified form of the cosmological constant could be reincorporated in our understanding of gravity to account for the dark energy.

An alternative explanation is suggested by the expectation from particle physics that completely empty regions of space can still produce a 'vacuum energy'. It is thought that pairs of particles and their associated anti-particles are continually created out of the vacuum, only to almost immediately annihilate each other and disappear. As they do so, they produce a minute outward pressure. Averaged over the entire voids of the universe, there would be sufficient such 'vacuum fluctuations' to produce enough pressure to push the universe further apart. Since this is a property of empty space, as the universe grows larger and the voids grow bigger, the effect of vacuum energy thus becomes increasingly dominant over that of gravity. However, a better understanding of particle physics is required before the vacuum energy theory can be shown to be a reality – current estimates suggest it would be much more powerful than observed even in our accelerating universe. Another rival for dark energy is the suggestion of a new force field which goes by the name of quintessence. Scientists are debating a rigorous mathematical description of quintessence, but it may be a force whose strength and importance changes through the history of the universe. It could be linked to the very early inflationary period of expansion thought to have occurred immediately after the initial Big Bang; if quintessence then lay relatively dormant for a period, we could be in another active phase producing the current accelerated expansion.

Whatever the true nature of the dark energy, if it continues to exert its influence we can speculate on a bleak future for our universe. The galaxies will continue to fly further and further apart ever faster, perhaps leading to an eventual 'Big Chill'. Even worse, if the effects of dark energy become increasingly important, it may begin to dominate over gravity on smaller scales, such as within galaxies – leading to a 'Big Rip'.

23

PLANETS, MOONS AND THE SEARCH FOR LIFE

The story of creation does not end with the birth of the stars. Indeed, for many other bodies in the universe, the Big Bang was just the start. Until some stars had been born, and eventually died, there could be no planets. And until there were planets there could be no life on Earth.

At the present time we have no proof that any form of life exists in the universe other than on our own planet. But before we start hunting for planets that may support life, we must understand how the Earth itself was formed. The first stars, formed hundreds of millions of years after the Big Bang, had no solid rocky planets orbiting around them. It was not until some of the stars had ended their lives with massive supernovae explosions that the space dust and the atoms of the heavier elements appeared in abundance throughout the galaxies. Only then could the rocky planets be created. A star and its accompanying planetary system will have formed from within a giant molecular cloud which eventually collapses under gravity, and after a few million years will reach pressures and temperatures sufficient to ignite nuclear fusion; a young star has formed at the core of the cloud. But the young star does not comprise all the material in the cloud. During its formation it is surrounded by a 'proto-planetary' disc that is also in the process of gravitational collapse subsequently to form the planetary system.

Individual planets form by a process called gravitational accretion. As the proto-planetary disc cools, particles of dust condense and form. As they pass close to each other, they are pulled together by their tiny gravitational attraction, until after a few millennia larger particles form. The larger lumps of matter – called planetesimals – are better able to attract additional particles and are thus more likely to grow than the smaller masses. The larger and more massive bodies continue to accumulate space debris and grow steadily. After more millennia of accumulation some have become a proto-planet, with a mass the size of a planet. Our solar system is an excellent example of planetary formation. Each early planetesimal had its own ring of space where it could gravitationally capture any smaller particle and grow a little larger. The exception is at the region of the asteroid belt, where the strong gravity of the proto-planet Jupiter disturbed the gravitational accumulation of the debris, preventing the formation of a single larger planet.

Solar Systems

After this evolutionary period, some of the stars had evolved planetary systems orbiting around them. Within these systems, planets known as gas giants – like Jupiter and Saturn in our own solar system – have been commonly identified. As yet, the smaller, rocky terrestrial planets are proving much harder to discover. Once a sizeable planet begins to form it can capture most of the matter within several million miles of its own orbit. Some of this matter itself condenses to form satellite moons around the proto-planet. We see very complex moon systems around all of the very massive gas giants but few in orbit around the inner rocky planets. The larger moons are true satellites that originated in the initial collapse; they are spherical in shape and they have orbits that are nearly circular. However, many other moons are smaller and irregularly shaped and often have very elongated orbits. They are not true satellites, but rather captured asteroids and comets that have succumbed to the gravitational field of the planet much later. Saturn has a spectacular ring system. Jupiter, Uranus and Neptune have fainter and less striking ring systems. Seen close up, the rings are found to be the remains of moons that have been broken into small pieces by the tidal forces of the planet.

How the Moon Was Formed

The Earth's moon (which we call 'the Moon') is very large compared with the size of our planet. It also has a much more complex and violent history than any of the other moons in the solar system. It is possible for a planet the size of the Earth to capture small objects to be held as satellites in orbit around it, such as we see in the case of the two small moons of Mars, but it is impossible for the Earth to capture a passing object as large as the Moon and to retain it in orbit. Over the years many theories have been suggested about the formation of the Moon. One suggested that the Moon had somehow broken off from the Earth, leaving a twin planetary system with both planets orbiting about their common centre of gravity. The truth is much more complex. At one time there were two planetesimals competing with each other for the matter in the space between Venus and Mars. The one destined to become the Earth was the larger and more successful, but the second planetesimal still managed to attract a substantial amount of the matter.

The two planetesimals both had elliptical orbits around the Sun. They avoided each other for millions of years, but then, billions of years before life began on the planet, a catastrophic collision took place between them. The impact was so great that the orbit of the Earth was considerably changed by the collision. The proto-Earth was greatly deformed, as the heat generated in the giant impact made the matter in the planetesimal molten and fluid, and a huge quantity of matter was thrown out into space into an orbit around the Earth. The other planetesimal disintegrated after the collision, except for its core, which had adhered to the proto-Earth during the collision.

After a time the two orbiting lumps of matter regained their spherical shapes and they evolved to become the Earth–Moon system. It happened that the Earth was very much the larger body. It is interesting to speculate what would have happened if the Moon had been larger and a twin planetary system had formed. Could this have created two life-supporting planets close to each other? What is not in doubt is that the tidal forces on the Earth would have been enormous, and a very different planet would have evolved. We know that the tidal forces of the Moon have played a major part in the evolution of life on our planet. It is also likely that the impact changed the Earth's axis of rotation to create the angle of the ecliptic – and therefore our familiar seasons.

The newborn Moon was so hot after the great collision that it remained molten for several millennia. As it cooled down, lakes of lava began to form on its surface and these eventually solidified to leave a crust. About four billion years ago the Earth and the Moon were subjected to a great barrage of debris from space, forming large craters on their surfaces. The heavily cratered lunar surface is testimony to this violent pounding. The Earth, however, although heavily scarred as well, was afforded some protection by the atmosphere and, over time, the weather has eroded the craters.

Are We Alone?

There is one major disappointment that has come about just because we now know so much about the solar system. Many different environments in the solar system have been discovered and explored, but there has been no positive sign of life anywhere other than on our own planet. There is one intriguing find in the form of the SNC meteorite discovered in Antarctica. It has been identified as a small piece of the planet Mars, thrown into space by a massive impact over a billion years ago with such velocity that it escaped the gravitational field of Mars and eventually landed on Earth. There is evidence to show that the rocky chunk had once been exposed to water and there is also evidence of fossilized primitive bacterial life, but scientists think the exposure to water was on Earth and not Mars. The *Mars Global Surveyor* and the Mars rovers *Spirit* and *Opportunity* have mapped the surface of Mars in detail. There is little doubt that at one time the planet had a much warmer climate and flowing water.

There is still a chance of finding primitive life forms, and the search for microscopic life continues. There has been speculation that some of the moons of the outer planets, in particular Saturn's moon Titan that is known to harbour complex organic compounds, could be suitable sites for life but they need a much warmer environment.

As early as 1952 those hoping to find evidence of life elsewhere in the solar system received encouragement when the American scientists Stanley Miller (1930–2007) and Harold Urey (1893–1981) performed a classic experiment with the simplest of laboratory equipment. They showed that in a closed container, using heat and electric sparks to simulate lightning, simple chemical elements such as hydrogen and nitrogen with molecules of water and carbon dioxide can combine to form organic molecules. Later experiments along the same lines have produced a wide variety of organic compounds. It is safe to conclude that the DNA molecule, and therefore life itself, could form under primitive Earth-like conditions.

New Directions

The space age has provided many new directions in which to take astronomy. The skies are being mapped in almost every frequency of the spectrum and in more detail than ever before. Objects once studied only through theory, such as black holes and planetary formations, are now the focus of intense observational study. We have close-up images of planetary surfaces, moons, and asteroids and comets on their journeys around the Sun. There are pulsars like the one in the Crab Nebula, the remains of an exploding supernova where the heavy elements are synthesized. There are countless galaxies from the nearby Andromeda Galaxy to distant active galaxies harbouring at their core black holes with a mass of over a billion suns. We can now observe so far back in the history of the universe that we will soon be able to see the first galaxies shortly after they formed, and we can map the distribution of galaxies throughout the firmament into the giant clusters and superclusters, themselves gathered to form the large-scale 'walls' around the void.

Humans have been on Earth for a few million years, and our recorded observations of the universe go back only a few thousand years. We have seen how humans have always looked with wonder at the skies. We have seen how the uncovering of the secrets of the universe has gradually taken place. This knowledge has been passed

THE STORY OF ASTRONOMY

on to future generations. We are the fortunate people that now inherit this knowledge, and we are able to appreciate the origins of the universe far better than our ancestors. We know that there is far more for us to discover in the skies, and that every generation will add new knowledge. It is important to remember also that we are the custodians of the Earth and we must look after it. For all we know, our fragile world may be unique in the universe.

GLOSSARY

Words in SMALL CAPITALS refer to other entries in the Glossary.

absolute magnitude The magnitude a star would have if it were located at ten PARSECS from the Earth.

accretion The gradual accumulation of matter by an astronomical body, usually by gravitation.

active galactic nucleus A GALACTIC NUCLEUS giving out strong emissions in the ELECTROMAGNETIC SPECTRUM.

active galaxy A very luminous GALAXY, usually containing an ACTIVE GALACTIC NUCLEUS.

aphelion The point of a planet's or comet's orbit at which it is farthest from the Sun. Opposite to PERIHELION.

apogee The point in the orbit of the Moon, or of any planet, at which it is at its greatest distance from the Earth; also the greatest distance of the Sun from the Earth.

apparent magnitude A measure of the brightness of a stellar object as seen from Earth.

asteroid A rocky object over a few hundred metres in diameter orbiting the Sun.

atmosphere The sphere of gases surrounding the Earth or any celestial body.

aurora Light radiated by atoms and IONS in the Earth's upper atmosphere.

azimuth An arc of the heavens extending from the ZENITH to the horizon, which cuts it at right angles; the quadrant of a great circle of the sphere passing through the zenith and NADIR.

barred spiral galaxy A SPIRAL GALAXY but with the spiral arms attached to a bar running through the nuclear bulge.

Big Bang The event that created the universe about 13 billion years ago, creating space, time, energy and matter.

binary star Sometimes called a double star. Two STARS that revolve around each other. They are held together by the force of their mutual GRAVITY.

black hole A body with such a strong gravitational field that light cannot escape from it.

blazar A type of ACTIVE GALAXY with very powerful emissions.

celestial object Any object visible in the night sky.

celestial sphere The whole of the night sky mapped onto a sphere.

cepheid variable star A pulsating yellow SUPERGIANT star used to calculate stellar distances.

cluster of galaxies A collection of a few hundred to a few thousand GALAXIES held together by their own gravity.

coma A spherical diffuse cloud of gas seen around the nucleus of a COMET near the Sun.

comet A small body of ice and dust in orbit around the Sun. The ice vaporizes near the sun giving rise to a characteristic tail.

conjunction The lining-up of three or more bodies. For example Earth–Venus–Sun, which gives rise to the transit of Venus across the Sun.

cosmic background radiation (CMR) The radiation from the primordial fireball known as the BIG BANG that fills all space.

cosmic ray High-speed particles travelling through space.

dark energy A repulsive gravitational effect causing the UNIVERSE to expand outwards.

dark matter Undetected missing matter from the universe with as yet unknown properties.

declination The angular distance of a heavenly body (north or south) from the celestial equator, measured on a meridian passing through the body. It corresponds to latitude on the Earth.

dwarf star Any star smaller than a giant, e.g. MAIN SEQUENCE stars and WHITE DWARFS.

eclipse The blocking of all (total eclipse) or part (partial eclipse) of the light from one celestial body by another.

ecliptic The plane of the Earth's orbit extended to infinity from the Sun. So called because eclipses can happen only when the Moon is on or very near this plane.

electromagnetic radiation A very wide range of radiation including gamma rays, X-rays, the optical spectrum, microwaves and radio waves.

electromagnetic spectrum The whole array of possible electromagnetic emissions.

electron An atomic particle with negative charge, usually found orbiting an atom.

elliptical galaxy A galaxy that is elliptical in shape with no spiral arms.

emission nebula A gaseous NEBULA, glowing by the light from a nearby star.

ether A medium believed to occupy the whole of space, carrying light as a wave motion.

Euclidean geometry The classical geometry of space as described by Euclid in the ancient world.

event horizon The boundary of a BLACK HOLE.

false-colour image An image, usually outside the optical part of the spectrum, showing the radiation in false colours.

galactic disc A disc of gas and dust surrounding the nucleus of a galaxy.

galactic nucleus The central part of a galaxy inside the nuclear bulge.

galaxy An assembly of young stars, gas and dust kept together by their mutual gravity.

gas giant A star with a radius of 10 to a 100 times that of the Sun.

geocentric The system of the world with the Earth as the centre.

gravitational energy The energy of a gravitational field such as is found around a BLACK HOLE.

gravity The property that all matter has an attraction for all other matter in the universe.

heliocentric The system of the universe with the Sun at the centre.

Hertzsprung–Russell (H–R) diagram A diagram plotting the absolute magnitude of stars against their surface temperature or spectral class.

Hubble constant (H_0) A constant relating the distance of a galaxy to its speed of recession. The reciprocal of H_0 determines the age of the universe.

infrared Radiation with a wavelength greater than the red end of the visible spectrum.

intercluster medium Gas and dust between the galaxies in a cluster.

intergalactic medium Gas and dust between neighbouring galaxies.

interplanetary medium The gas and dust in the space between the planets.

ion An atom becomes an ion when it loses one or more valence ELECTRONS and thus acquires a positive charge, or gains one or more electrons and thus acquires a negative charge.

kelvin (K) Temperatures measured from absolute zero and referred to in units of kelvin, after Lord Kelvin (1824–1907).

Kuiper Belt A ring of space beyond the planets that is the birthplace of COMETS in the SOLAR SYSTEM.

lenticular galaxy A disc-shaped galaxy, but without the spiral arms.

light year The distance travelled by light in a year. About 5.88×10^{12} miles (9.46×10^{12} km).

local group The MILKY WAY and its neighbouring galaxies.

magnitude A measure of the brightness of any object in the sky. Plotted on a logarithmic scale with the brighter stars as the lower magnitudes.

main sequence A grouping of stars on the HERTZSPRUNG–
RUSSELL DIAGRAM showing the hottest and brightest to
the coolest and dimmest stars.

magellenic clouds Two neighbouring galaxies seen from
the Southern Hemisphere.

mass The mass is the amount of matter in a body. (This
tautology was first put forward by Isaac Newton.)

Messier catalogue The catalogue compiled by Charles
Messier (1730–1817) consisting of all the nebulous
objects in the sky.

meteor A streak of light seen in the sky when any space
debris is vaporized by the Earth's atmosphere.

meteorite A fragment of space debris that survives the
atmosphere to land on Earth.

meteroid A fragment of space debris orbiting the Sun.

microwaves ELECTROMAGNETIC RADIATION with wavelengths
of the order 0.001 to 1 metre.

Milky Way The galaxy to which the Sun belongs.

nadir The point of the sky directly below the point of
observation. The opposite to a ZENITH.

nebula A cloud of interstellar gas and dust.

neutron An uncharged particle found in the nucleus of
an atom.

neutron star A very dense remnant of a collapsed star,
consisting almost entirely of NEUTRONS.

nova A new star. Usually a star in a binary system which suddenly emits powerful radiant energy.

nuclear fusion Atomic energy generated by the fusing together of atomic nuclei.

Oort Cloud A hollow spherical shell far out in the SOLAR SYSTEM, mainly populated by comets.

orbit The path of an astronomical object moving round another such object.

parallax The apparent displacement, or difference in the apparent position, of an object, caused by actual change (or difference) of the position of the point of observation. In astronomy there are two kinds of parallax, viz diurnal and annual, the former when a celestial object is observed from opposite points on the Earth's surface, the latter when observed from opposite points of the Earth's orbit. The horizontal parallax is the diurnal parallax of a heavenly body seen on the horizon.

parsec A measure of distance based on PARALLAX; 3.26 light years.

perihelion The point of a planet's or comet's orbit at which it is nearest to the Sun. Opposite to APHELION.

photon A particle of light.

photosphere The region in the Sun's atmosphere where most of the visible light originates.

planetary nebula A luminous shell of gas ejected from an old low-mass star.

plasma Hot ionized gas (see ION).

precession (of the equinoxes) The earlier occurrence of the equinoxes in each successive sidereal year, due to the retrograde motion of the equinoctial points along the ecliptic, produced by the slow change of direction in space of the Earth's axis.

proton A positively charged particle found in the nucleus of an atom.

protoplanetary disc A disc of material surrounding a PROTOSTAR or a new star where planets could be created.

protostar A sphere of gas still growing in mass until it is dense enough to shine and become a star.

pulsar A pulsating radio source emitted from a rotating NEUTRON STAR.

quasar A powerful 'quasi-stellar radio source' with a very large REDSHIFT.

radio galaxy A galaxy that emits most of its radiation in the radio part of the SPECTRUM.

radio waves Long wavelength ELECTROMAGNETIC RADIATION, used for radio and TV communication.

radius vector A variable line drawn to a curve from a fixed point as origin: in astronomy the origin is usually at the Sun or a planet round which a satellite revolves.

red dwarf A low-mass star on the main sequence of the H–R DIAGRAM.

red giant A large cool star with a high luminosity.

redshift The shifting of light to longer wavelengths. It can be caused by gravitational fields or by high speeds of recession. The shifting of light to shorter wavelengths is called blueshift.

reflection nebula A dense cloud of gas in interstellar space, illuminated by the light of stars lying behind it.

relativity The theory of time and space in the universe as developed by Albert Einstein.

satellite A body in orbit around a larger astronomical body.

Seyfert galaxy A SPIRAL GALAXY with a bright nucleus and with emission lines in its spectrum.

solar system The system of all the astronomical objects subjected to the Sun's gravity.

solar wind A flow of PROTONS and ELECTRONS emitted by the Sun.

spectrum The result of passing light through a prism to spread out the various colours. The spreading out of other parts of ELECTROMAGNETIC RADIATION by similar means.

spiral arms Arms of gas, dust and stars associated with SPIRAL GALAXIES.

spiral galaxy A flattened and rotating galaxy with two spiral arms winding out from the nuclear bulge at the centre of the galaxy.

star A sphere of gas giving out radiation in the ELECTRO-MAGNETIC SPECTRUM.

starburst A place where there is an exceptionally high rate of star formation.

stellar wind Equivalent to SOLAR WIND, but applied to any star.

supergiant A star with very high luminosity.

supernova A rare stellar explosion when a star can increase its brightness by a millionfold.

supernova remnant The remnant left over when a SUPER-NOVA loses its brightness.

transit The passage of one astronomical body across the face of another, for example the planet Venus on the face of the Sun.

ultraviolet The part of the ELECTROMAGNETIC SPECTRUM beyond the violet end of visible light.

universe The whole of space, consisting of matter, energy and radiation.

visible light The part of the ELECTROMAGNETIC SPECTRUM that can be seen with the naked eye.

white dwarf A stellar remnant that has exhausted all its thermonuclear fuel and can no longer shine.

X-ray radiation High-frequency radiation between the gamma ray spectrum and ULTRAVIOLET light.

X-ray star A NEUTRON STAR in a binary system that emits bursts of X-rays.

zenith The point of the sky directly overhead. The highest point of the celestial sphere viewed from any particular place. The opposite to a NADIR.

zodiac The 12 sections of the sky recognized by astronomers and astrologers. Astrologers lay great store by which planets appear in which signs of the zodiac.

FURTHER READING

Armitage, A., *Edmond Halley* (Nelson 1966)

Aughton, P., *Newton's Apple* (Weidenfeld & Nicolson 2002)

Aughton, P., *Transit of Venus* (Weidenfeld & Nicolson 2004)

Buttman, G., *William Herschel* (WVG Stuttgart 1961)

Comins, N.F. and Kaufmann, W.J., *Discovering the Universe* (W.H. Freeman 2005)

Donahue, W.H. *Johannes Kepler's New Astronomy* (Cambridge University Press 1992)

Gribbin, J., *In Search of the Big Bang* (Heinemann 1986)

Hall, A.S., *The Scientific Revolution 1500–1800* (Longman 1954)

Hawking, S., *A Brief History of Time* (Bantam 1988)

Hawking, S., *The Universe in a Nutshell* (Transworld 2001)

Hoffmann, B., *Einstein* (Paladin 1975)

Hogben, L., *Science for the Citizen* (George Allen 1938)

Longair, M.S., *Our Evolving Universe* (Cambridge University Press 1996)

Mitton, S., *The Cambridge Encyclopaedia of Astronomy* (Jonathan Cape 1977)

Moore, P., *Watchers of the Stars* (Michael Joseph 1974)

Roos, M., *Introduction to Cosmology* (Wiley 2003)

Sagan, C., *Cosmos* (Abacus 1980)

Skeat, W.W., *Chaucer, the Complete Works* (Oxford University Press 1912)

Weinberg, S., *The First Three Minutes* (André Deutsch 1977)

Westfall, R.S., *Never at Rest* (Cambridge University Press 1980)

White and Gribbin, *Stephen Hawking, A Life in Science* (Penguin 1992)

Wolf, A., *A History of Science, Technology and Philosophy* 2nd edn (Allen Unwin 1950)

Wright, W.A. (ed), *The Rubaiyat of Omar Khayyam* (Macmillan 1973)

INDEX

Abell, George, 351
Abul Wala, 57
Académie des Inscriptions et Belles-Lettres, 147
Académie Royale d'Architecture, 147–8
Académie Royale des Sciences, 147–8, 149–50
active optics, 321
Adams, John Couch, 192–3
adaptive optics, 321
Adelard of Bath, 65–6
al-Birini, 57–8
al-Hakim, Caliph, 59
Al Sufi Abd al-Rahman (Azophi), 56–7
Albategnius, 56, 60, 61
Alexander the Great, 24
Alexandria (Egypt), 24, 36, 37–8, 53
  Lighthouse, 35
algebra, 55, 61–2
Alhazen, 58–9

Allen Telescope Array, 274–5
Almagest, 43–9, 67
amateur astronomers: role, 325
Andalusia, 53–4, 65–6
Andromeda Galaxy, 57, 233–4, 237–8
Anglo-Saxon Chronicle, 68
angular measurement: origin of degrees and minutes, 18
Antikythera instrument, 51–2
Apollo missions, 306–7
Apollonius of Perga, 45, 48, 62, 93
Arabian and Persian astronomy, 53–66
arc of the meridian, 159
Archimedes, 98
Aristarchus, 25–31
Aristotle, 24–5, 32, 105–6
Armstrong, Neil, 306–7
Arrest, Heinrich Louis d', 193
Arzachel, 59–60
astrolabes, 51–2, 54–5, 60, 70–1, 162

astrology
    Arab, 57–8
    beginnings, 12
    in Renaissance, 72, 73, 82–3, 84,
        87, 88, 94–5
astrophotography, 203–5
atmospheric refraction, 34
atomic weapons, 225–6
atoms, 198, 247–58, 265–9
Augustus, Roman emperor, 42
Azophi see Al Sufi Abd al-Rahman
Aztecs, 13–14

B²FH paper, 265–7, 297–8
Babbage, Charles, 188
Babylonians, 18–20
Barberini, Cardinal see Urban VIII,
        Pope
Barrow, Isaac, 135
Bell, Jocelyn, 261–3
Ben Sabbah, 60–1
Bessel, Friedrich W., 200
Besso, Michele, 211
Betelgeuse, 252–3
Big Bang
    and black holes, 302
    criticisms, 243, 265–6
    Hubble's contribution, 240–1
    origins of theory, 222
    overview, 326–39
    and singularities, 299
    situation, 289
black holes
    first suggestion of existence,
        155–6
    overview, 269–73, 276–9,
        298–303
    and quasars, 281, 298
    supermassive, 352–3

theory universe is contained
    inside, 287–9
Bohr, Niels, 248–9
Bondi, Hermann, 243, 297
Brahe, Tycho, 81–8, 90, 95, 116,
    258
Brown, Michael, 318
brown dwarfs, 349
Bruno, Giordano, 70, 96–7
Brunowski, Johann, 91
Burbidge, Geoffrey, 298
Burbidge, Margaret, 298
Byzantines, 52–3

Caccini, Tommaso, 103
calculus, 133–4, 156–7, 189
    tensor calculus, 224
calendars
    ancient, 13–14
    Arab, 61
    Babylonian, 18–20
    Chinese, 49–50
    Egyptian, 20–1
    Gregorian, 41, 56, 73–4
    Julian, 41–2
    Sumerian, 17–18
Callisto, 314, 315
Cambridge University, 112–14,
    129–30, 135, 261–2, 292–3
Camden, William, 110
Canopus, 33–4
carbon: formation, 253, 266–7
Cassini, Gian-Domenico, 151–2,
    158
Cassini, Jacques, 159
Cassini, Jean-Dominique, comte de,
    160
Cassini de Thury (Cassini III),
    159–60

Cassini spacecraft, 312, 316
Cassini division, 158–9
Castelli, Benedetto, 104
Centurion (ship), 173
Cepheid variables, 230–1, 237–8, 341
Ceres, 318
CERN, 333
Chaldeans, 33
Challis, James, 192–3
Chandra Observatory and Telescope, 322, 324
Chandrasekhar, Subrahmanyan, 256, 324
Charles II, king of Great Britain and Ireland, 165
Charon, 318
Chaucer, Geoffrey, 70–1
Cheever, Ezekiel, 114
chemical compounds: notation system, 198
Chinese astronomy, 12, 49–50, 258
Christian IV, king of Denmark, 87
Chwolson, Orest, 223
El Cid, 65
clocks and watches
    17th-century manufacture, 111
    at Greenwich, 167
    radium clocks, 249
    regulation by pendulum, 96–7, 99
    shipboard, 172–6
CMB (cosmic microwave background), 327–30, 335–6, 339
Colbert, Jean-Baptiste, 147, 151
collimators, 322
Columbus, Christopher, 162
comets
    and amateur astronomers, 325
    discovered by SOHO, 323
    Halley's, 68–9
    Helvelius' observations, 126–7
    Kuiper Belt, 319
    mentions in *Anglo-Saxon Chronicle*, 68
    Oort Cloud, 319
    Shoemaker-Levy 9, 313, 325
    supposed influence on life on Earth, 11, 49
    Tempel 1's contrived collision with spacecraft, 324–5
compasses, 194
Compton Telescope, 322
conic sections, 48, 62, 93
Cook, Captain James, 176–7
Copernicus, Nicolaus
    influence, 83, 91, 96–7, 104, 116
    influences on, 60
    overview, 71–80
Cosmic Background Explorer, 329
cosmic microwave background *see* CMB
cosmological constant, 221–2, 355–6
cosmology
    ancient Greek, 23, 25–31, 37, 44–9
    Arab, 56
    Babylonian, 19–20
    Copernican, 71–80
    Galileo's support for Copernican, 104–8
Crab Nebula, 263
Crabtree, William, 95, 116, 118, 119, 121–2
cubic equations, 62

Curtis, Heber, 233–4
Cygnus X-1, 273

Dalton, John, 198
dark energy, 340, 355–7
dark matter, 340, 348–51, 352–3
Darwin, Charles, 201–2
Davy, Sir Humphrey, 196
Dawn Mission probe, 318
Deep Impact spacecraft, 324–5
deferents and epicycles system,
    44–5
Deptford, HMS, 174
Desargues, Girard, 149
Descartes, René, 148, 149, 150
Dicke, Robert, 328
Digges, Captain Dudley, 174–5
dimensions: fourth, 283–6
Din Malik Shah, sultan of Jalal, 61
Discovery space shuttle, 319
Doppler, Christian, 232
Doppler effect, 232
Draper, Henry, 204
Draper, John, 204
dust, 282–3, 346
dwarf planets, 318

$E = mc^2$, 221, 226, 247, 252, 338,
    355
Earnshaw, Thomas, 176
Earth
  age, 201–3
  Arab cosmological theories, 56
  Copernican cosmology, 71–80,
    104–8
  craters, 363
  distance from Sun, 152, 177
  early cosmological theories, 13,
    19–20, 23, 25–31, 37, 44–9

  death, 253
  escape velocity, 270–1
  formation, 358–61
  and formation of the Moon,
    361–3
  measuring, 31–2, 151
  orbit, 56, 59
  rotation cycles, 40
  shape, 159
Easter, calculating date, 74
eclipses, lunar
  ancients' knowledge, 26, 28,
    39–40, 44, 47
  and longitude calculation, 163
  supposed influence on life on
    Earth, 82
eclipses, solar
  ancients' knowledge, 5–7, 23,
    25–6, 37–8, 44, 47
  and longitude calculation, 163
  mentions in Anglo-Saxon Chronicle,
    68
Eddington, Sir Arthur, 220, 223,
    252, 341
Egyptian astronomy, 11, 20–1
Einstein, Albert, 136, 209–27, 232
Einstein, Eduard, 211
Einstein, Elsa, 225
Einstein, Hans Albert, 211
Einstein, Mileva, 211
Einstein Cross, 223–4
Einstein Ring, 223
electricity, 194–7
  nuclear-generated, 251
electromagnetism, 224, 333, 337
electrons, 248
Electroweak Era, 337
electroweak force, 333, 337
elements

discovery, 198–9
formation, 253, 254, 256–7,
    265–9, 297–8
entropy, 298, 299
epicycles *see* deferents and epicycles
    system
equinoxes, precession of, 40, 42, 47
equivalence, principle of, 217–21
Eratosthenes, 31–2
Eris, 318
escape velocities, 155–6, 270–1
ether, 206–9
Euclid, 62
Eudoxus, 45
Euler, Leonhard, 157, 176
Europa, 314, 315
EUVE (Extreme Ultraviolet
    Explorer), 322–3
Evans, Revd Robert Owen, 325
Evelyn, John, 136–8
event horizons, 271, 277
evolution theory, 201–2
extraterrestrial life, 96–7, 261–3,
    274–5, 358, 363–4
Extreme Ultraviolet Explorer *see*
    EUVE

Far Ultraviolet Spectroscopic
    Explorer *see* FUSE
Faraday, Michael, 195–6, 226–7
Fermat, Pierre de, 149
Fisher, Richard, 342
Fitzgerald, Edward, 63
FitzGerald, George, 209
Flamsteed, John, 125, 152, 166–9,
    170–1
force concept, 332–4
Fowler, William, 298
France: maps, 159–60

Frederick II, king of Denmark, 62,
    84–5, 86, 87
Friedmann, Alexander, 221–2, 326
FUSE (Far Ultraviolet Spectroscopic
    Explorer), 322–3

Gagarin, Yuri, 305–6
galaxies
    classification, 238–9
    clustering, 351–2
    measuring distance from Earth,
        344–5, 351–2
    movements, 239–43
    rotation and structure, 346–53
    *see also individual galaxies by
        name*
Galileo spacecraft, 312, 313–14
Galileo Galilei, 96, 97–108, 148,
    149, 164, 291
Galle, Johann Gottfried, 193
gamma rays, 322
Gamow, George, 326–7
Ganymede, 314–15
gas giants, 314, 360
Gassendi, Pierre, 123, 148
Geller, Margaret, 352
Gellibrand, Herbert, 116, 119
geology, 202–3, 249
geometry, 57, 62
George III, king of Great Britain
    and Ireland, 184
George, Prince, of Denmark, 169
Gervaise of Canterbury, 69–70
Gold, Thomas, 243, 297
Graham, George, 172
Grand Unification Epoch *see* GUE
gravitational accretion, 359
gravitational lensing, 222–4,
    350–1

gravity
  and black holes, 269–70, 271–2,
    277–8
  Copernican attitude, 75
  and dark energy, 355–7
  escape velocities, 155–6, 270–1
  Galileo's experiments, 99–100
  importance, 193–5
  measuring specific, 98
  Newton's theory, 131–3, 134,
    135–6, 138–45
  and planetary orbits, 158
  separation from other forces
    during Big Bang, 334
  and space/time, 216–20
  and unified field theory, 333–4
'Great Debate', 233–4
Great Wall, 352
Greek astronomy, 22–40, 43–9
Green, Charles, 177
Greenwich Observatory, 151, 165–9
Gregory XIII, Pope, 41
Grossmann, Marcel, 224
GUE (Grand Unification Epoch),
    335
Guth, Alan, 335

H-R diagram, 341–2
Habicht, Konrad, 212
Hadron and Lepton Era, 337–8
Hale, George Ellery, 234–5, 236
Halley, Edmond, 68–9, 138–43,
    169–71
Harrison, John, 171–6
Harrison, William, 174
Harvard, John, 114
Harvard Observatory, 229
Harvard University, 114
Hawking, Jane, 293, 294

Hawking, Stephen, 278, 290–304
Heisenberg, Werner, 250
helium, 338
Herodotus, 20–1, 22
Herschel, Caroline, 180–1, 182,
    184, 185, 189
Herschel, John, 126, 184, 186,
    188–91
Herschel, Margaret, 190
Herschel, Mary, 185, 188
Herschel, William, 178–89
Hertzsprung, Ejnar, 341
Herzog, Albert, 211
Hesiod, 22
Hevelius, Elizabeth Margarethe,
    127
Hevelius, Johannes, 126–7
Hipparchus, 36–40, 44, 46, 47
Hooke, Robert, 138–40, 153
Horace, 78
Horrocks, Jeremiah, 95, 110–26,
    168
Hoyle, Fred, 243, 253, 265–7, 292,
    296–8
Hubble, Edwin, 222, 234, 235,
    236–42, 344
Hubble Space Telescope, 237, 241,
    317, 319–22, 345
Hubble's constant, 239–42, 344–5
Hubble's law, 239–42, 344, 354–5
Huchra, John, 224, 352
Huchra Lens, 224
Humason, Milton, 239, 344
Huygens, Christiaan, 138, 316
Hven, 84–6
hydrogen, 338
hydrogen line, 342–3, 346–7
hydrogen spectrum, 199
hydrostatic balances, 98

Inflationary Epoch, 335–6
infrared, 187
interferometry, 321
Io, 314
Islamic astronomy *see* Arabian and
    Persian astronomy
Israel, Werner, 303

James I and VI, king of England
    and Scotland, 85
Jansky, Karl, 259–60
Jeans, Sir James, 243, 297
Jodrell Bank, 261
Julius Caesar, 41, 42
Jupiter
    ancients' understanding of
        movements, 11
    and the asteroid belt, 360
    and Copernican cosmology,
        78
    exploration, 312–16
    moons, 101, 314–16
    orbit, 158
    rings, 360–1
    use to calculate speed of light,
        152–3
    use to determine longitude, 151,
        163–4

Keck Telescopes, 321
Kelvin, Lord, 202, 246–7
Kendall, Larcum, 176
Kennedy, John F., 306
Kepler, Johannes, 81, 87–95, 116,
    258
Kepler's Star, 91
Khwajah Nizami, 64
Kuiper, Gerard, 319
Kuiper Belt, 319

Lagrange, Joseph-Louis, 157, 158,
    188
Laika (dog), 305
Lansberg, Philip, 116
Laplace, Pierre Simon, 157, 158,
    188
Large Hadron Collider, 333
latitude and longitude, 36, 151–2,
    159, 161–77
Le Verrier, Urbain-Jean-Joseph,
    193, 219–20
leap years, 41
Leavitt, Henrietta, 229–31, 236, 341
Leibniz, Gottfried, 156
Lemaître, Georges, 222, 326
Levy, David, 325
light
    Herschel's experiments, 187
    Newton's experiments, 130–1
    speed of, 152–3, 215–16
    waves or particles?, 153–4,
        206–9, 249–50
    see also optics
light years, 200
Lippershey, Hans, 100
Liverpool, 109–11
longitude *see* latitude and longitude
Lorentz, Hendrick, 209
Lorini, Nicolo, 103
Louis XIV, king of France, 150–1
Lowell, Percival, 308
Lowell Observatory, 231
lunar parallax, 46, 341
Lyell, Charles, 202–3

magnetism, 194–7
    see also electromagnetism
maps, 159–60
Mariner missions, 307, 311

Mars
  ability to support life, 363–4
  ancients' understanding of
    movements, 11
  and Copernican cosmology, 77,
    78
  exploration, 307–10
  Galileo's observations, 102
  and Horrocks' hypothesis, 123
  orbit, 92–3
  use to determine Earth–Sun
    distance, 152
Maskelyne, Nevil, 173, 182
Mather, Richard, 112
matrix mechanics, 250
Maxwell, James Clerk, 196–7, 224,
  226–7
Mayans, 14
Mayer, Tobias, 176
Mayer, Walter, 225
Mendeleyev, Dmitry, 199
Mercury
  ancients' understanding of
    movements, 11
  and Copernican cosmology, 77,
    78
  death, 253
  exploration, 311–12
  Gassendi's observations of transit,
    123
  orbit, 219
Mersenne, Marin, 148–9
Messenger mission, 312
Messier, Charles, 182
meteors, 11, 69–70
Michell, John, 156
Michelson, Albert, 207–9, 215
microwaves, 197
Milky Way

Herschel's model, 186–7
  position in universe, 241
  rotation, 347–8
  size, 236
  structure, 235–6, 346–9
  Sun's position, 186–7, 235–6, 346
Miller, Stanley, 364
Miranda, 317
Mohammed, Prophet, 53
Moon
  ancients' understanding of
    movements, 2, 5–10, 13, 26–9,
    44, 46
  craters, 363
  dark side, 306
  distance from Earth and size,
    26–9, 37, 46
  Earth's gravitational force on,
    135–6
  first examination by telescope,
    101
  formation, 361–3
  landings, 306–7
  libration, 157
  lunar parallax, 46, 341
  maps, 126–7, 181
  meteorites striking, 69–70
  orbit, 124–5
  phases, 9–10
  photographs, 204
  rotation, 159
  and Sumerians, 17
  and tides, 10
  use to determine longitude, 151,
    163, 164–5, 167–9, 175–7
  worship, 14
  see also eclipses, lunar
moons: formation, 360–1
Morley, Edward, 208, 215

motion, Newton's laws, 133–4,
    143–5
Mount Wilson Observatory, 234–7
Much Hoole, 117–18
mural quadrants, 86

navigation
    arc of the meridian, 159
    astrolabes, 51–2, 54–5, 60, 70–1,
        162
    compasses, 194
    measuring longitude, 151,
        161–77
    nautical almanacs, 170–7
    origin of degrees and minutes, 18
    origin of latitude and longitude,
        36
nebulae, 181, 182, 204, 244–5, 346
    planetary, 254
    spiral, 231–4, 237–8
Neptune, 192–3, 312, 317, 360–1
neutron stars, 257, 264–5
neutrons, 337
New Horizons probe, 318
Newton, Humphrey, 145–6
Newton, Isaac
    date of birth, 291
    on Descartes, 150
    and Flamsteed's research, 169
    on Horrocks, 126
    and light, 153–4
    as Master of the Mint, 191
    and Moon's motion, 168
    overview, 128–46
    rivalry with Leibniz, 156
    and telescopes, 131, 180
    validity of laws, 226–7
Nizam ul Mulk, 60–1
Novara, Domenico Maria de, 72

nuclear fission, 251
nuclear force
    strong, 332, 333, 335
    weak, 332, 333
nuclear fusion, 251
nuclear physics, 246–58
numerical systems: zero, 55

observatories
    Greenwich, 151, 165–9
    Harvard, 229
    Hubble Space Telescope, 237,
        241, 317, 319–22, 345
    Lowell, 231
    Mount Wilson, 234–7
    Palomar, 235, 237
    Paris, 150–2
    in space, 322–4
    Uraniborg, 84–6
Odoacer, 52
Oenopides, 23
Olber, Heinrich, 155
Olber's paradox, 155
Omar Khayyam, 60–5
Oort, Jan, 319
Oort Cloud, 319
Opportunity rover, 309
optics, 58–9
    active and adaptive, 321
Orion Nebula, 204

pair production, 335
Palomar Observatory, 235
Pandora, 316
parallax method, 340–1
    lunar parallax, 46, 341
    solar parallax, 123–4
    stellar parallax, 200–1, 204–5,
        341

Parbsjerg, Manderup, 82
Paris Observatory, 150–2
parsecs, 200
Pascal, Blaise, 149
Pascal's triangle, 62
Peacock, George, 188–9
Penrose, Roger, 278, 299–300, 302–3
Penzias, Arno, 328
periodic table, 199
Perlmutter, Saul, 354
Persian astronomy *see* Arabian and Persian astronomy
PG 2112+059 (quasar), 282
Phoenix lander, 309
photoelectric effect, 213–14
photography *see* astrophotography
photons, 208, 248–9, 338–9
pi, 18–19
Picard, Jean, 151–2
Pickering, Edward, 205, 229
Planck, Max, 330
Planck Era, 334
Planck scale, 330–1
plane of the ecliptic, 48
planetesimals, 359
planets
    ancients' understanding of movements, 10–11, 44–5, 46, 47–9
    and Copernican cosmology, 76–9
    Descartes' theories on motions, 150
    dwarf planets, 318
    escape velocities, 155–6, 270–1
    formation, 358–61
    gas giants, 314, 360
    gravity's effect on orbits, 138–45, 158

Horrocks' hypothesis, 123
Kepler's laws of planetary motion, 93–4
life on other, 96–7, 261–3, 274–5, 358, 363–4
measuring distance from Earth, 341
orbits and positions, 79, 81–95
see also individual planets by name
Pluto, 318
Poseidonius, 33–4
prime mover, 48
prisms, 130–1
probability, binomial, 62
Prometheus, 316
protons, 337
protostars, 245, 339
Ptolemy
    influence, 67, 72, 75, 83, 91
    and the Moon, 125
    on orbit shape, 93
    overview, 43–9
pulsars, 261–5
Pythagoras, 23

quadrants, mural, 86
quantum mechanics, 225, 250, 296, 300–2
quarks, 337
quasars, 276, 279–83, 298
quintessence, 356–7

radio waves, 197
ratios, 62
Reber, Grote, 260
Reconnaissance orbiter, 309
red giants, 252–8, 266
redshift, 231–2, 280–1, 344–5, 352

relativity
  and black holes, 277
  general theory of, 216–21, 222–4
  and quantum mechanics, 296,
    300–2
  special theory of, 208–9, 210,
    213–16
Rhaeticus, 80
Rhodes, Colossus of, 35–6
ring systems, 360–1
Roman empire, fall of, 52
Rømer, Ole, 152–3
Romulus Augustus, Western
    Roman emperor, 52
rotation curve, 348
Royal Astronomical Society, 188
Royal Society, 131, 138, 147,
    149–50
RR Lyrae variables, 236
Rubin, Vera, 348
Rudolf II, Holy Roman Emperor,
    87, 90, 94–5
Rudolphine Tables, 90–1, 94–5, 116
Russell, Henry Norris, 341
Rutherford, Ernest, 247–8

Sagredo, Giovanni Francesco, 105
Salviati, Fillipo, 104
Sandage, Allan, 279
satellites, manmade, 305, 306
Saturn
  ancients' understanding of
    movements, 11
  and Copernican cosmology, 78
  exploration, 312, 316–17
  moons, 159, 185–6, 316–17
  orbit, 158
  rings, 101–2, 158–9, 316, 360
Schiaparelli, Giovanni, 308

Schmidt, Brian, 354
Schmidt, Maarten, 280
Schrödinger, Erwin, 250
Schwarzchild radius, 272, 287
Scout missions, 309–10
Secchi, Angelo, 205–6
SETI (Search for Extraterrestrial
    Intelligence), 274–5
Seven Wonders of the Ancient
    World, 35–6
Shapley, Harlow, 233–4, 235–6
Shoemaker-Levy 9 (comet), 313,
    325
singularity theory, 299–300, 302
Sirius, 11, 254
Slipher, Vesto, 231–2, 239, 344
SNC meteorite, 363–4
SOHO (Solar and Heliospheric
    Observatory) spacecraft, 323
solar parallax, 123–4
Solovine, Maurice, 212
solstices, precession of, 42
Sosigenes, 41
South, James, 189–90
space exploration, 305–25
spectrographs, 205
spectroscopy, 205–6, 231–3
spiral nebulae, 231–4, 237–8
Spirit rover, 309
Spitzer Space Telescope, 322–3
Sputnik I, 305
stars
  ancients' understanding of
    movements, 47, 48
  binary, 181–2, 189–91, 273
  birth, 245–6, 339, 346
  death, 252–8
  distribution in sky, 186–7,
    189–91

and element formation, 253, 254,
    256–7, 265–9, 297–8
escape velocities, 271
globular star clusters, 235–6
Helvelius' maps, 127
magnitudes and positions, 36–7,
    57, 86–7, 200–1, 204–5,
    229–31, 341–4
neutron stars, 257, 264–5
photographs, 204
spectral features, 205–6, 231–3
stellar parallax, 200–1, 204–5,
    341
supernovae, 83–4, 91, 257–8,
    261–9, 325
telescopes lead to discovery of
    many more, 102–3
variable stars, 230–1, 236, 237–8,
    342
why they shine, 245–7, 251–2
see also individual stars by name
steady-state theory, 243, 297,
    327–9
stellar parallax, 200–1, 204–5, 341
Strabo, 33
Stukeley, William, 132
Sumerians, 16–18
Sun
    age, 202, 246–7
    ancients' understanding of
        movements, 2, 5–10, 13, 23,
        26, 44, 48
    and Babylonians, 19
    classification as star, 154
    and Copernican cosmology,
        71–80, 104–8
    death, 252–4
    distance from Earth and size,
        29–30, 38–40, 46, 152, 177

distance from planets, 94
gravitational effect on planets'
    orbits, 138–45
position in Milky Way, 186–7,
    235–6, 346
solar parallax, 123–4
solar wind, 323
space observation of, 323
why it shines, 251–2
worship, 14
see also eclipses, solar; stars
supernovae, 83–4, 91, 257–8,
    261–9, 325
Type 1a, 343–4, 354
super-sphere, 285–6

Tariq ibn Ziyad, 66
telescopes
    Allen Telescope Array, 274–5
    Chandra, 322
    Compton, 322
    examples through history, 2
    Hale, 237
    Herschel's, 179–81, 185–6
    Horrocks' poem about, 115
    Hubble Space, 237, 241, 317,
        319–22, 345
    invention, 100–1
    Jodrell Bank, 261
    Keck Telescopes, 321
    Mount Wilson, 234, 235
    new techniques, 321
    radio, 259–63
    reflecting, 131, 179–81
    Spitzer Space, 322–3
Tempel 1 (comet), 324–5
tensor calculus, 224
Thales, 23
thermodynamics, 298–9

3C 48 (star), 279
3C 273 (star), 279, 280
tides, 10
time
  and black holes, 277, 278
  and special theory of relativity,
    213, 215
time measurement
  on board ship, 172–6
  and distance from equator,
    169–70
  length of months, 42
  and longitude calculation, 163–76
  origin of days of the week, 19
  origin of minutes and seconds,
    18
  see also calendars; clocks and
    watches
Titan, 316–17, 364
Toledo, 59, 65–6
Tombaugh, Clyde, 318
Towneley, Richard, 152
trigonometry, 57
Triton, 317
Tully, Brent, 342
21-cm hydrogen line, 342–3, 346–7
Twin Quasar, 223

umbra, 28
uncertainty principle, 250
unified field theory, 333–7
universe
  age, 154, 241–2, 344–5
  expansion, 221–2, 231–2,
    239–43, 353–7
  'Great Debate', 233–4
  infinity of, 154–5, 353–5
  measuring, 340–9
  nature of, 285–9

size, 187–8, 200–1, 231–4,
    237–9
  steady-state theory, 243, 297,
    327–9
  structure, 344–52
Uraniborg, 84–6
uranium, 249
Uranus, 182–4, 312, 317, 360–1
Urban VIII, Pope (Cardinal
    Barberini), 104, 105, 106
Urey, Harold, 364

vacuum energy theory, 356
Venera probes, 310
Venus
  ancients' understanding of
    movements, 10
  and Copernican cosmology, 77,
    78
  death, 253
  exploration, 310–11
  orbit and size, 119–23
  phases, 102
  use to determine Earth–Sun
    distance, 177
Viking missions, 307–8
virtual particles, 300–1
voids, 352
Vostok I, 305–6
Voyager probes, 312, 316, 317

Wallis, John, 112–13, 156
watches see clocks and watches
water hole, 274
Watzenrode, Lucas, 72, 73
wave equation, 250
Whiston, William, 170–1
white dwarfs, 254
Wickens, John, 129

Wilkinson Microwave Anisotropy probe, 329
Willis, Roger, 173
Winteler, Jost, 211
wormholes, 279
Worthington, John, 113, 119
Wren, Christopher, 138–40, 165–6

X-rays, 197, 273, 322, 324

zero, 55
zodiac constellations, 12
Zwicky, Fritz, 223, 349